MOLECULAR STRUCTURE AND BONDING

The Qualitative Molecular Orbital Approach

MOLECULAR STRUCTURE AND BONDING

The Qualitative Molecular Orbital Approach

BENJAMIN M. GIMARC

Department of Chemistry
University of South Carolina
Columbia, South Carolina

 1979

ACADEMIC PRESS   New York   San Francisco   London

A Subsidiary of Harcourt Brace Jovanovich, Publishers

ACADEMIC PRESS, INC.
111 Fifth Avenue, New York, New York 10003

United Kingdom Edition published by
ACADEMIC PRESS, INC. (LONDON) LTD.
24/28 Oval Road, London NW1 7DX

Library of Congress Cataloging in Publication Data

Gimarc, Benjamin M
    Molecular structure and bonding.

    Includes bibliographical references.
    1. Molecular structure. 2. Chemical bonds.
I. Ttile.
QD461.G56   541'.22   78–3336
ISBN 0–12–284150–6

PRINTED IN THE UNITED STATES OF AMERICA

79 80 81 82   9 8 7 6 5 4 3 2 1

Contents

Preface

The quantitative successes of molecular orbital calculations are well known, but the qualitative aspects of MO theory may not be so widely appreciated. In this book I hope to show that the remarkably simple and often powerful qualitative principles of MO theory lead to some very appealing interpretations of molecular properties and provide a conceptual framework for chemistry. Surveys of broad classes of compounds can point out gaps in our experimental knowledge and suggest problems for more rigorous theoretical methods.

I have tried to make the book understandable to readers whose knowledge of chemistry is at the senior undergraduate or first-year graduate level. I have assumed a familiarity with quantum chemistry which most students at that level might reasonably be expected to have achieved. No knowledge of formal group theory is required but the reader should have an appreciation of molecular symmetry such as those ideas usually covered in most recent undergraduate physical chemistry textbooks. Beyond the pedagogical value of teaching the elements of simple molecular orbital theory and then using those principles to unify a sizable body of chemistry, I hope the conclusions and predicted trends presented in this book will stimulate further research by those professionals who do molecular structure studies, synthetic chemistry, thermodynamics, and ab initio quantum mechanical calculations of molecular electronic structures.

In Chapter 1 I have tried to record the rules of qualitative MO theory, but I found it difficult to commit to paper all the intuitive rules one uses unconsciously. Justifications or rationalizations accompany each rule and usually a simple example illustrates the idea. The heart of the book, Chapters 2–8, deals with applications that are centered on qualitatively deduced MO diagrams. Chapter 9 reviews attempts to provide a more rigorous understanding of, and justification for, the success of qualitative MO theory. Chapter 10 describes the historical evolution of the model and mentions directions for future development and applications. References appear at the end of each chapter. Where large numbers of chemical species are surveyed, the references are arranged by molecule. In such cases the references to a specific molecule may be a source of structural data (bond distances and bond angles or perhaps only the point group or shape), but if that information is not available the references will contain reports of other properties or possibly only a claim that the molecule has been detected. Often the most reliable structural data for small reactive molecules come from ab initio quantum mechanical calculations.

Acknowledgments

I have many people and organizations to thank for their assistance, inspiration, encouragement, and support in the preparation of this book. My research, on which much of this book is based, has been supported by the National Science Foundation and the University of South Carolina. Much credit goes to the graduate students and my faculty colleagues at the University of South Carolina for their stimulating and often unreasonable questions that forced me to start thinking about the qualitative aspects of MO theory. I wrote the book while on sabbatical leave at Priceton University, and I must thank the Chemistry Department of that institution for accommodating me during a most pleasant and productive year. I am particularly grateful to Professor Leland C. Allen of Princeton, whose hospitality and generosity made possible my stay at Princeton. My conversations with Professor Allen and his critique of the rough draft of this book had a profound effect on its outcome, but in the end I must claim complete responsibility for what appears on the following pages. Finally, I thank my wife, Jerry Dell, for her patience, understanding, and emotional support.

1 The rules of qualitative molecular orbital theory

INTRODUCTION

Chemical properties are determined by the laws of physics. What makes chemistry different from physics is the limitless number of molecules that can be formed from various combinations of a hundred or so different kinds of atoms. The fascination and frustration of chemistry comes from having so many examples. A detailed knowledge of the properties of one molecule does not in itself give us any knowledge about the properties of other molecules. What we need in chemistry, in addition to specific details about individual molecules, is an understanding of how properties differ among and within broad classes of molecules.

The most successful approach for the quantitative application of quantum mechanics to chemistry has been through the molecular orbital approximation. A hierarchy of quantitative MO procedures exists: the ab initio Hartree–Fock method with configuration interaction, limited basis set ab initio methods such as the STO-3G level, semiempirical SCF MO methods such as the various CNDO and MINDO approximations, and semiempirical, non-SCF schemes such as the extended Hückel method and pi-electron Hückel theory. A review of nearly twenty years of accumulated results of rigorous ab initio SCF MO calculations at or near the Hartree–Fock limit leads one to conclude that most of the properties of chemistry related to the electronic structure of molecules can be accounted for within the molecular orbital approximation (1).

Although quantitative theory is an essential part of science, most chemists do not accept the agreement between numerical calculations and experimental results as a sufficient explanation of molecular properties. They demand, in addition, a qualitative rationalization of why the numerical results turn out as they do. The experimental or calculated data must be fit together or synthesized by means of a qualitative model to make the trends among the data comprehendable.

Traditional or conventional qualitative valence theory has grown out of the valence bond approximation as developed by Pauling. This theory has a rich heritage that can be traced back through the work of Heitler and London to the models of Lewis and Langmuir. Molecular properties are explained in terms of bond pairs and lone pairs of valence electrons, resonance structures, and electrostatic repulsions among localized electron pairs.

Molecular orbital theory offers an alternative qualitative model for understanding molecular properties. It is aesthetically pleasing to have an interpretive model that grows directly out of the most powerful and convenient quantitative method of application of the fundamental principles that underlie chemistry. The qualitative MO model should be able to provide a means for guided speculation about molecular properties that could lead to new experiments or detailed quantum mechanical calculations. Because the qualitative arguments fit within the MO framework, they can be checked by any of several different levels of MO calculations. MO theory can treat both ground and excited states of molecules. Finally, the MO model is simple and general enough to play an important role in chemical education.

Although much of this book deals with molecular shapes, that is, the gross features of molecular geometry, these considerations lead quite naturally to the study of other properties including barriers to internal rotation and inversion, hydrogen bridging, hydrogen bonding, reaction mechanisms, relative molecular stabilities, and trends in bond strengths, bond distances, and bond angles through families or classes of related molecules.

QUANTUM MECHANICS AND MOLECULAR ORBITAL THEORY

The application of quantum mechanics to chemistry requires the approximate solution of Schrödinger's wave equation,

$$H\psi = W\psi, \qquad [1]$$

where the eigenvalue W is the total energy of the system under study and the eigenfunction ψ the wavefunction. If the wavefunction ψ is normalized to unity,

$$\int \psi^2 \, dv = 1,$$

then ψ^2 can be interpreted physically as an electron probability density function. Assuming fixed nuclei, the Hamiltonian operator H is given by

$$H = \sum_i T(i) + \sum_{i,\mu} V_{ne}(i, \mu) + \sum_{i<j} V_{ee}(i, j) + \sum_{\mu<\nu} V_{nn}(\mu, \nu), \qquad [2]$$

where $T(i)$ are the kinetic energy operators for individual electrons i, $V_{ne}(i, \mu)$ Coulombic attractions between electrons i and nuclei μ, $V_{ee}(i, j)$ the electron-electron repulsion operators, and $V_{nn}(\mu, \nu)$ the nuclear-nuclear repulsions. The two-electron terms $V_{ee}(i, j)$ make the wave equation [1] difficult to solve because they link together the coordinates of all the electrons into a system of multidimensional partial differential equations that cannot, in general, be solved.

Since ψ and W are unknown, the usual practice is to approximate ψ by a function Φ of known form and then calculate an approximate energy E as

$$E = \int \Phi H \Phi \, dv \bigg/ \int \Phi^2 \, dv.$$

The function Φ can be chosen on the basis of physical interpretability and computational convenience. The variation theorem guarantees that for an approximate function Φ, the calculated approximate energy E is an upper bound to the true energy W, or $E \geqslant W$. A particularly meaningful and useful form for the choice of Φ is the molecular orbital wavefunction,

$$\Phi_{MO} = \phi_1(1)\phi_1(2) \cdots \phi_n(2n),$$

expressed as a product of molecular orbitals ϕ_i that are functions of the coordinates of individual electrons. The MOs ϕ_i distribute individual electrons throughout the entire nuclear framework of the molecule rather than localizing them in particular bonds or on atoms. The Pauli exclusion principle stipulates that no more than two electrons can be assigned to the same ϕ_i. More accurately, the Pauli principle is satisfied by using a function Φ that is the antisymmetrized product of MOs or by taking the Slater determinant of which Φ_{MO} above is the main diagonal product.

The approximate total energy E is minimized by a process called the self-consistent-field (SCF) method, which results in the expression

$$E = \sum \varepsilon_i - V_{ee} + V_{nn}, \qquad [3]$$

where ε_i are orbital energies associated with individual MOs ϕ_i and the summation is over all electrons. The quantities V_{ee} and V_{nn} are the total potential energies of electron-electron and nuclear-nuclear Coulombic repulsions.

Although Φ_{MO} is not a solution to the Schrödinger wave equation [1], it is a solution to a simpler problem. Suppose we could approximate the true Hamiltonian operator H by a sum of effective one-electron operators $H_{eff}(i)$,

$$H \sim H' = \sum_i H_{eff}(i),$$

the sum being over all electrons i. An obvious way to form H' is by neglecting entirely the electron-electron repulsion terms $V_{ee}(i, j)$ in H in Eq. [2] and then grouping the remaining one-electron kinetic energy operators $T(i)$ and the nuclear-electron attraction terms $V_{ne}(i, \mu)$ together as $H_{eff}(i)$. This drastic approximation is not satisfactory because the electron repulsion part of the total energy is not small. Instead, one might hope to choose $H_{eff}(i)$ so as to account for the effect of interelectronic repulsions in some average way.

The MO wavefunction Φ_{MO} is an eigenfunction of the approximate Hamiltonian H',

$$H'\Phi_{MO} = E'\Phi_{MO},$$
[4]

because each term $H_{eff}(i)$ contains only the coordinates of electron i and therefore $H_{eff}(i)$ operates only on an individual MO $\phi(i)$. This permits the many-electron wave equation [4] to be separated into identical, independent, one-electron eigenvalue equations,

$$H_{eff}(i)\phi_l(i) = \varepsilon_l \phi_l(i),$$
[5]

where ε_l is the energy of an electron in orbital ϕ_l. Then

$$E' = \sum \varepsilon_l,$$
[6]

where the sum is over all electrons. The complexity of the situation is therefore greatly reduced by substituting for the original problem [1] a simpler and, one might hope, related problem [5].

Suppose we approximate ϕ_l by a linear combination of atomic orbitals (AOs) χ_r from the various atoms that compose the molecule,

$$\phi_l = \sum_r c_{rl} \chi_r.$$
[7]

This is called the LCAO MO (linear combination of atomic orbitals) approximation. Suppose a molecule is composed of m atoms and each atom contributes one or more AOs for a total of n AOs. The collection of n AOs $\{\chi_r\}$ is called the basis set, from which a set of n MOs $\{\phi_l\}$ can be constructed.

For convenience and without any loss of generality we require the AOs to be normalized,

$$\int \chi_r^2 \, dv = 1,$$

but not necessarily orthogonal, so that in general

$$\int \chi_r \chi_s \, dv = S_{rs} \neq 0,$$

where S_{rs} is the overlap between the two AOs χ_r and χ_s. Using the LCAO expansion of Eq. [7] to calculate an approximate orbital energy ε_l gives

$$\varepsilon_l = \frac{\int \phi_l H_{eff} \phi_l \, dv}{\int \phi_l^2 \, dv} = \frac{\sum_{r,s} c_{rl} c_{sl} \int \chi_r H_{eff} \chi_s \, dv}{\sum_{r,s} c_{rl} c_{sl} \int \chi_r \chi_s \, dv} = \frac{\sum_{r,s} c_{rl} c_{sl} H_{rs}}{\sum_{r,s} c_{rl} c_{sl} S_{rs}},$$

where S_{rs} is the overlap integral above and

$$H_{rs} = \int \chi_r H_{eff} \chi_s \, dv.$$

Applying the variation theorem to ε_l, each coefficient c_{rl} can be varied to minimize the orbital energy ε_l, which leads to a set of n homogeneous, linear equations,

$$\sum_r^n c_{rl}(H_{rs} - \varepsilon_l S_{rs}) = 0, \qquad s = 1, 2, \ldots, n. \qquad [8]$$

These equations have nontrivial solutions for the coefficients c_{rl} only if the $n \times n$ determinant formed from the quantities in the parentheses in [8] is zero:

$$|H_{rs} - \varepsilon S_{rs}| = 0. \qquad [9]$$

Equation [9] is the well-known secular equation or secular determinant. If the quantities H_{rs} and S_{rs} were known, the secular determinant could be solved for n values of ε_l, which could then be substituted back into the homogeneous linear equations [8] to obtain n sets of coefficients $\{c_{rl}\}$, a different set for each ϕ_l and corresponding ε_l.

Slater-type AOs may be used as χ_r to evaluate the overlap integrals S_{rs}. Tables and formulas are readily available (2). It is possible to approximate H_{rs} semiempirically, using atomic spectral data and other assumptions, and avoiding the explicit specification of $H_{eff}(i)$. The integral H_{rr} can be interpreted as being related to the energy of an electron in an AO χ_r and, therefore, H_{rr} can be approximated by the experimental ionization potential (IP) of an electron in the AO χ_r on the atom involved. Tables of experimental valence state IPs are available (*3*). Assuming that χ_r and χ_s are on different atoms, as they usually are, we can argue that H_{rs} ($r \neq s$) should be some kind of interaction energy between atoms r and s. If the distance between the two atoms is large, then the interaction H_{rs} is small. At distances approximating those of chemical bonds, H_{rs} might be related to the energy of a bond between the two atoms. A very simple choice is to make H_{rs} directly proportional to the overlap S_{rs}: $H_{rs} = KS_{rs}$, where K is a number to be chosen to make calculated results agree with those from experiment. Since S_{rs} is normally positive and H_{rs} is a bond energy, lower or more stable than the energy of the separated atoms, K must be negative. Other prescriptions in which H_{rs} is formed by averaging H_{rr} and H_{ss} have also been used (*4*).

With specified values for H_{rs} and S_{rs}, the secular equation [9] can be solved for the n orbital energies ε_l, which can be substituted into the homogeneous linear equations [8] for the n sets of coefficients $\{c_{rl}\}$ of the n MOs ϕ_l. The total energy E' is just the sum of orbital energies $\sum \varepsilon_l$, where the summation is over all electrons with no more than two electrons in the same MO.

Since molecular geometry enters the computation through the overlap integrals S_{rs}, it is possible to do calculations for different sets of atomic coordinates to learn how the quantities ϕ_l, ε_l, and E' change for various molecular conformations. Computer programs are widely available to do just this kind of MO calculation. The scheme was developed by Roald Hoffmann who named it the extended Hückel method (*4*). One of the objectives of this book is to show that the qualitative conclusions of such calculations can often be obtained without actually carrying out the calculations. Because of the highly approximate nature of extended Hückel calculations, these qualitative conclusions are usually the only results of much value. In other cases the results of extended Hückel or more rigorous MO calculations can be interpreted or rationalized qualitatively.

Before proceeding further we need to reconsider three total energy quantities: the true energy W from the exact solution of the wave equation [1], the approximate SCF MO energy E of Eq. [3], and the orbital energy sum E' from Eq. [6]. Experience shows that the MO energy E differs from the true energy W by about 1%, an amount that is large compared with energy quantities of chemical interest such as barriers to internal rotation, bond energies, and heats of reactions. For many chemical processes, E fortunately parallels W quite closely and there is no reason to expect it to do otherwise so long as Φ_{MO} remains a good approximation to the true wavefunction ψ throughout the change. The relationship between E' on the one hand and E and W on the other is not apparent. Equation [3] shows that the MO energy E is not simply a sum of orbital energies ε_l. Even if the orbital energies ε_l of Eq. [5] approximate the ε_l in Eq. [3], the difference

$V_{nn} - V_{ee}$ in E is not negligible. Furthermore, E is calculated using the correct Hamiltonian H, but E' is associated with H' for another problem.

A complete discussion of these difficulties is deferred to Chapter 9, which reviews the important work that ultimately justifies the use of the orbital energy sum E' as a measure of the changes in the true energy during variations in molecular shape. Although there is a large difference between E' and E, E' does change in very nearly the same way as E during variations in molecular shapes.

ORTHONORMALITY OF THE MOs

The MOs are required to be orthonormal:

$$\int \phi_l^2 \, dv = 1, \qquad \int \phi_k \phi_l = 0.$$

Consider the details of the normalization integral:

$$\int \phi_l^2 \, dv = \int \left(\sum_r c_{rl} \chi_r \right) \left(\sum_s c_{sl} \chi_s \right) dv = \sum_r c_{rl}^2 + 2 \sum_{r<s} c_{rl} c_{sl} S_{rs} = 1.$$

The sums of squares of coefficients plus cross terms multiplied by overlaps add up to unity. Whether the coefficients are positive or negative, the squared terms are always positive and the overlaps S_{rs} are also usually positive. If all the coefficients have the same sign, the summation to unity requires that the numerical values of the individual coefficients be less than unity. If the MO contains nodes or changes of phase between atoms, then some of the cross terms will be negative. For the normalization sum to equal unity in these cases, the squared terms will have to be larger than those in the nodeless case for them to be able to cancel the negative cross terms and have the sum equal to unity. Higher energy MOs have more nodes or phase changes and thus they have larger coefficients.

CHARGE DENSITIES AND BOND ORDERS

The square of a normalized MO ϕ_l can be interpreted as a probability density function for one electron. Integration of ϕ_l^2 over all space gives a total probability of unity for finding the electron:

$$\sum_r c_{rl}^2 + 2 \sum_{r<s} c_{rl} c_{sl} S_{rs} = 1.$$

In order to simplify AO indexing, assume that each MO includes only one AO from each atom. This will often be true as a result of molecular symmetry. The squared terms c_{rl}^2 are related to the amount of charge on each atom r and the cross terms $2c_{rl}c_{sl}S_{rs}$ measure the amount of charge in the region between atoms r and s from the MO ϕ_l.

Mulliken (5) has defined the overlap population or bond order in terms of the cross terms. Let $p_{rs}(l)$ be the overlap population between atoms r and s in MO ϕ_l:

$$p_{rs}(l) = 2n_l c_{rl} c_{sl} S_{rs},$$

where n_l is the number of electrons occupying ϕ_l ($n_l = 0, 1,$ or 2). The total overlap population or bond order p_{rs} is the sum over all MOs ϕ_l,

$$p_{rs} = \sum_l p_{rs}(l).$$

Large bond orders can be correlated with short bond distances. Therefore, bond orders can be used to account for gross differences between single, double, and triple bonds and even to rationalize trends in bond distances for similar bonds through a series of related molecules.

It might seem appropriate to define the atomic populations or charge densities as simply the squared terms c_{rl}^2, but in order to include all of the electron charge, Mulliken chose to include part of the overlap population as well, dividing it equally between atoms r and s. Therefore, let $q_r(l)$ be the gross atomic population on atom r due to MO ϕ_l:

$$q_r(l) = n_l \left[c_{rl}^2 + \sum_{r<s} c_{rl} c_{sl} S_{rs} \right].$$

The total gross atomic population or charge density is the sum of $q_r(l)$ for all MOs:

$$q_r = \sum_l q_r(l).$$

THE SYMMETRY OF MOLECULAR ORBITALS

Molecular orbitals must be either symmetric or antisymmetric with respect to the symmetry operations of the molecule. This follows directly from the electron density interpretation of the electronic wavefunction and the electron indistinguishability principle of quantum mechanics. These symmetry restrictions severely limit the number and kinds of AOs that can combine in a particular MO, thus simplifying the job of forming MOs since most small molecules have high symmetry.

In order to use qualitative MO theory, the only part of group theory required is a knowledge of symmetry classifications. Some general rules are helpful. MOs of linear molecules are denoted by the symbol σ if they are cylindrically symmetric about the molecular axis. The π MOs have a plane of antisymmetry that contains the molecular axis. For nonlinear molecules, energetically discrete or nondegenerate MOs are designated by the letter a if they are symmetric with respect to rotation about the principal n-fold axis. Antisymmetric orbitals are designated by the letter b. There can obviously be no MOs that are antisymmetric with respect to n-fold rotations if n is odd. If a molecule has a principal rotational axis of 3-fold or higher, there can be doubly degenerate MOs and these are labeled e. The e MOs of structures with odd-fold principal axes can be neither symmetric nor antisymmetric with respect to the odd-fold rotations. Tetrahedral and octahedral molecules can have triply degenerate MOs that are labeled t.

Orbitals can be further classified as being either symmetric (') or antisymmetric (") with respect to reflection in a plane of symmetry. In other cases planar reflection symmetry is indicated by subscript 1 (symmetric) or subscript 2 (antisymmetric). The D_{nh} groups ($n \geqslant 3$) have 2-fold axes that are perpendicular to the principal n-fold axis. The MOs are labeled as being either symmetric (subscript 1) or antisymmetric (subscript 2) with respect to these perpendicular 2-fold axes. In the D_2 and D_{2h} groups, subscripts 1, 2, and 3 indicate symmetry with respect to one of three mutually perpendicular 2-fold axes while being antisymmetric with respect to the other two. Finally, if a molecule contains a center of inversion, the MOs are classified as being either symmetric (subscript g) or antisymmetric (subscript u) with respect to inversion through that center. Many examples follow in the chapters on applications.

The hybrid AOs of conventional valence theory do not necessarily have the symmetry of the molecule and, therefore, they are not taken as starting AO combinations in the application of MO theory. For example, the oxygen atom lone pair sp^3 hybrids in the water molecule are individually asymmetric with respect to the 2-fold rotational axis and the molecular plane of reflection. On the other hand, the nitrogen lone pair hybrid in ammonia is symmetric with respect to the 3-fold rotational axis and the three vertical reflection planes that contain the 3-fold axis. Therefore, we might expect to find something like a hybrid orbital resulting from the MO treatment of ammonia but it is not necessary to assume AO hybridization a priori.

It is easier to compose MOs for structures of higher symmetry. The orbital degeneracies that result from high symmetry guarantee equal energies of two or more orbitals, reducing the number of different orbital energies to be estimated. High symmetry also limits the kinds of combinations of AOs that can enter a particular MO. Orbitals for less symmetric shapes can then be generated by angular deformations of the MOs for higher symmetry. In some cases the lower symmetry MOs so produced may not contain the full complement of AOs allowed by the lower symmetry, but such deficient MOs can be corrected by mixing pairs of MOs of the same symmetry classification. It is this orbital mixing that produces hybridlike orbitals in MO theory. Mixing is more important for higher energy MOs and for orbitals of similar energies. For the purposes of this book we can satisfactorily account for mixing in the higher energy MOs by mixing only the MO pair of highest energy for a given symmetry classification, accepting as adequate representations those lower energy orbitals produced directly by angular variations of the higher symmetry structure. Furthermore, the highest energy pair need be mixed only if the two MOs are composed of different kinds of AOs. An important result of this MO mixing is an added energy lowering or stabilization of the lower energy orbital and a destabilization of the higher orbital.

In later chapters we study changes in MO energies that occur as molecular shapes vary or as molecules symmetrically dissociate into fragments or as monomers dimerize symmetrically. In some instances it will seem that MOs of the same symmetry should logically cross as a result of these changes. This cannot happen, however, because of the noncrossing rule: Orbitals of the same symmetry cannot cross. Instead, they mix and diverge. This is an example of the mixing of orbitals of similar energies.

MOs FOR H_2

In order to illustrate a few rules concerning the relative energies and nodal properties of MOs, orbital symmetry, and the relative sizes of AO coefficients, we perform MO calculations for some simple molecules.

The AO basis set for H_2 consists of two $1s$ AOs, one on each hydrogen. Call these s_A and s_B. From two AOs it is possible to construct two MOs as linear combinations of the AOs. In the lower energy MO the AOs combine with the same sign or phase to form a bonding MO σ_g (1):

$$\sigma_g = c(s_A + s_B).$$

1

The two hydrogen atoms are exactly equivalent and the MOs must be either symmetric or antisymmetric with respect to interchange of the two hydrogens. Therefore, the coefficients of the two AOs in σ_g must be equal. The out-of-phase or antibonding combination is σ_u (2):

$$\sigma_u = d(s_A - s_B).$$

2

The antisymmetric combination σ_u has AOs with coefficients of equal magnitude but opposite sign.

The high symmetry of H_2 allows us to determine the coefficients c and d from the MO normalization property alone. Requiring that σ_g be normalized, we obtain

$$1 = \int \sigma_g{}^2 \, dv = c^2 \int (s_A + s_B)^2 \, dv = c^2 \int (s_A{}^2 + 2s_A s_B + s_B{}^2) \, dv$$

$$= c^2(1 + 2S_{AB} + 1) = 2(1 + S_{AB})c^2,$$

since the AOs are themselves normalized. Letting $S_{AB} = S$,

$$c = [2(1 + S)]^{-1/2}.$$

For two $1s$ AOs on hydrogens separated by the bond distance in H_2 (0.7 Å), the overlap $S = 0.75$. Substituting this value into the equation above for c yields $c = 0.54$. Requiring that σ_u be normalized produces

$$d = [2(1 - S)]^{-1/2} = 1.41.$$

Collecting results:

$$\sigma_g = 0.54(s_A + s_B), \qquad \sigma_u = 1.41(s_A - s_B).$$

Notice that MO normalization requires that the AO coefficients in σ_u be larger than those in σ_g. Nodal surfaces introduce phase differences between AOs in a MO. As a general rule these phase differences and the normalization requirement produce large AO coefficients.

The bond population of σ_g occupied by two electrons ($n = 2$) is

$$p_{AB}(\sigma_g) = 2 \times 2c^2 S = 0.88.$$

Of course, σ_u is empty in H_2, but if it were doubly occupied its bond population would be

$$p_{AB}(\sigma_u) = 2 \times 2d(-d)S = -6.0,$$

the minus sign indicating the antibonding character of σ_u. Notice that σ_u is more antibonding than σ_g is bonding. Quite generally, the antibonding interactions of out-of-phase AO combinations outweigh the bonding interactions in comparable in-phase combinations of the same AOs.

The secular equation for H_2 is

$$\begin{vmatrix} H_{AA} - \varepsilon S_{AA} & H_{AB} - \varepsilon S_{AB} \\ H_{BA} - \varepsilon S_{BA} & H_{BB} - \varepsilon S_{BB} \end{vmatrix} = 0.$$

From symmetry $H_{AA} = H_{BB} = \alpha$, $H_{AB} = H_{BA} = \beta$, $S_{AB} = S_{BA} = S$. From AO normalization $S_{AA} = S_{BB} = 1$. The secular equation reduces to

$$\begin{vmatrix} \alpha - \varepsilon & \beta - S\varepsilon \\ \beta - S\varepsilon & \alpha - \varepsilon \end{vmatrix} = 0.$$

Expanding this determinant gives

$$(\alpha - \varepsilon)^2 - (\beta - S\varepsilon)^2 = 0,$$

which on rearrangement to standard form can be solved by the quadratic formula for two values of the orbital energies ε:

$$\varepsilon_g = \frac{\alpha + \beta}{1 + S}, \qquad \varepsilon_u = \frac{\alpha - \beta}{1 - S}.$$

Since α and β are inherently negative quantities, the lower energy is ε_g.

Let ΔE_g and ΔE_u be the energy differences between the ε_g and ε_u and the energy α of an isolated hydrogen atom:

$$\Delta E_g = \alpha - \frac{\alpha + \beta}{1 + S} = \frac{\alpha S - \beta}{1 + S}, \qquad \Delta E_u = \frac{\alpha - \beta}{1 - S} - \alpha = \frac{\alpha S - \beta}{1 - S}.$$

The numerators of ΔE_u and ΔE_g are identical. Because S is positive,

$$\Delta E_g = \frac{\alpha S - \beta}{1 + S} < \frac{\alpha S - \beta}{1 - S} = \Delta E_u,$$

or the stabilization ΔE_g of σ_g is less than the destabilization ΔE_u of σ_u (Fig. 1). The result here demonstrates the energetic manifestation of the difference in bond orders: If overlap is included, the energy of the out-of-phase combination is always raised more than the energy of the comparable in-phase combination is lowered.

The quantitative differences between the results for H_2 and those for similar diatomics such as H_2^+ and He_2 depend only on the choice of a specific value for the overlap S. The qualitative conclusions are the same. For example, the ratio of coefficients in σ_u and σ_g, $d/c = [(1 + S)/(1 - S)]^{1/2}$, which must be a number greater than unity so long as $0 < S < 1$. Thus, whether we consider H_2, H_2^+, or He_2, σ_u must be more antibonding than

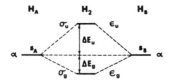

FIG. 1   Comparison of orbital energy levels (ε_g and ε_u) of the H_2 molecule with those (α) of the separated H atoms.

σ_g is bonding. Similarly, the ratio of orbital energy changes relative to the isolated atoms is $\Delta E_u/\Delta E_g = (1 + S)/(1 - S)$, which is also greater than unity. The energy of σ_u is raised above that of the isolated atom more than the energy of σ_g is lowered. H_2^+ has only one electron in σ_g and, therefore, the energy lowering relative to separated atoms is ΔE_g instead of $2 \Delta E_g$ for H_2. We conclude therefore that the bond energy in H_2^+ should be about half that in H_2, in rough agreement with experimental values of 61 kcal/mole for H_2^+ and 103 kcal/mole for H_2. In He_2, with its four electrons filling both σ_g and σ_u, there would be a net destabilization or energy increase relative to the separated atoms. Therefore, He_2 does not exist as a stable molecule. We are using here a principle of great value to qualitative MO theory: Molecules with the same numbers of similar kinds of atoms are readily compared in MO theory because the atoms contribute the same numbers and kinds of AOs from which qualitatively similar sets of MOs can be constructed. Individual molecules in such a class differ primarily in the number of valence electrons that occupy the common MO system.

Figure 1 compares the energies of the σ_g and σ_u MOs of H_2 with the energies of the separated AOs. How do the MO energies compare with that of the united atom He? Suppose the two hydrogen atoms in H_2 are forced together to form He. As the distance between the two nuclei decreases, the overlap S between the two $1s$ AOs increases smoothly to unity, causing the orbital energy ε_g to reach a minimum at zero internuclear distance. It is clear that orbital energies, calculated from Eq. [4], do not contain the internuclear repulsions which prohibit the collapse of H_2 into He. Furthermore, in our simple MO model, the fusion of two He atoms to make Be is prevented, not by increased internuclear repulsions, but because σ_u is more antibonding than σ_g is bonding. In general, the orbital energy sum $E' = \sum \varepsilon_l$ is part of the electronic component of the total energy, and in qualitative MO theory it is the only quantity we consider. Although internuclear repulsions are conceptually simple, there is no satisfactory way to include them qualitatively with the orbital energy sum E'. This book is mainly concerned with energy changes that result from angular geometry variations or from chemical reactions. Since these are situations in which atoms do not approach each other more closely than normal bond distances, the neglect of internuclear repulsions rarely causes the theory to break down.

MOs FOR H_3^+

The ion H_3^+ is known to have equilateral triangular geometry. The AO basis set consists of three $1s$ AOs, s_A, s_B, s_C, centered on protons H_A, H_B and H_C (3). Solving the secular equation for triangular H_3^+ gives three MO

3

energies, the upper pair being degenerate as required by the 3-fold symmetry (Fig. 2). The lower energy MO a_1' is the in-phase overlapping combination of the three $1s$ AOs (**4**), appearing with the same coefficient as required by symmetry;

$$a_1' = c(s_A + s_B + s_C).$$

4

For $1s$ AOs separated by 0.9 Å, their approximate distance in $H_3{}^+$, the overlap is 0.7. Requiring a_1' to be normalized,

$$\int (a_1')^2 \, dv = c^2(3 + 6S) = 1,$$

which on substituting $S = 0.7$ yields $c = 0.37$ and $a_1' = 0.37(s_A + s_B + s_C)$.

Each of the degenerate e' MOs has a node that cuts through the center of the triangle, the node in one component of e' being perpendicular to the node in the other. Otherwise, the location or orientation of the nodes of degenerate orbitals is arbitrary. Therefore, for simplicity we can require that the node of e_y' bisect the triangle to eliminate s_C from this orbital. By thus choosing e_y' to be antisymmetric we require that s_A and s_B appear with coefficients of equal magnitude but opposite sign (**5**):

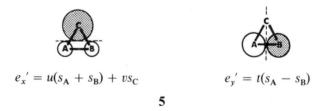

$$e_x' = u(s_A + s_B) + vs_C \qquad\qquad e_y' = t(s_A - s_B)$$

5

Since the node in e_x' is perpendicular to that in e_y', the coefficients of s_A and s_B in e_x' must be equal but the weighting of s_C may be different. The larger circle for s_C anticipates a larger coefficient for that AO. All three coefficients

$$e' \quad \underline{\qquad} \qquad \epsilon(e') = \frac{\alpha + \beta}{1 + S}$$

FIG. 2   Energy levels for triangular $H_3{}^+$.

$$a_1' \quad \underline{\qquad} \qquad \epsilon(a_1') = \frac{\alpha + 2\beta}{1 + 2S}$$

u, v, t of the e' orbitals can be determined by orthonormalization requirements. For e_x' to be orthogonal to a_1':

$$\int a_1' e_x' \, dv = 0; \qquad c \int (s_A + s_B + s_C)[u(s_A + s_B) + v s_C] \, dv = 0,$$

$$2(1 + 2S)u + (1 + 2S)v = 2u + v = 0, \qquad v = -2u.$$

Absolute values of u and v can be obtained by normalizing e_x':

$$\int (e_x')^2 \, dv = 2u^2(1 + S) + 4uvS + v^2 = 1.$$

Substitution of $S = 0.7$ and $v = -2u$ leads to

$$(3.4 - 5.6 + 4)u^2 = 1, \qquad u^2 = 1/1.8, \qquad u = 0.75, \qquad v = -2u = -1.50.$$

Normalizing e_y', we obtain

$$\int (e_y')^2 \, dv = 2t^2(1 - S) = 1.$$

Substituting $S = 0.7$ yields $t = 1.29$. Collecting results, we have

$$a_1' = 0.37(s_A + s_B + s_C),$$

$$e_x' = 0.75(s_A + s_B) - 1.50 s_C,$$

$$e_y' = 1.29(s_A - s_B).$$

For triangular H_3^+, symmetry considerations and MO orthonormality requirements suffice to determine the coefficients. Notice that nodes give the e' MOs larger coefficients than a_1'. Orthogonalization of e_x' against a_1' produces a balance between regions of opposite phase in e_x'; the coefficient of the lone s_C AO being larger than that of the in-phase pair s_A and s_B. These are further examples of how nodes or phase changes within a MO influence the size of the AO coefficients. The relative size of AO coefficients in the e' orbitals of H_3^+ has an importance that goes far beyond this example. They are the same as those of the hydrogen $1s$ AO coefficients in the e orbitals of ammonia and of the CH_3 groups in ethane.

PERTURBATION MO THEORY

The perturbation treatment of molecular orbital theory has been highly developed and widely applied (6). For the development of rules for qualitative MO theory we need to appropriate only the simplest results from PMO theory. Some of the terminology of perturbation theory is particularly expressive and, therefore, useful in qualitative arguments.

Imagine two independent systems A and B. These could be either a pair of atoms or a pair of molecules or one of each. System A has an orbital ϕ_A with an energy level ε_A and associated with system B are ϕ_B and ε_B. Furthermore, suppose that there is a large energy difference between A and B, with A having the lower energy $\varepsilon_A \ll \varepsilon_B$. Now introduce a perturbation interaction between A and B by letting h' be a perturbing operator in the Hamiltonian of A + B. The perturbation treatment to second order, neglecting overlap for simplicity, is exactly equivalent to solving the secular equation

$$\begin{vmatrix} \varepsilon_A - \varepsilon & \beta \\ \beta & \varepsilon_B - \varepsilon \end{vmatrix} = 0,$$

where

$$\beta = \int \phi_A h' \phi_B \, dv.$$

Expanding the determinant leads to the quadratic equation in ε,

$$\varepsilon^2 - (\varepsilon_A + \varepsilon_B)\varepsilon + \varepsilon_A \varepsilon_B - \beta^2 = 0,$$

which can be solved for two roots using the quadratic formula:

$$\varepsilon = \tfrac{1}{2}(\varepsilon_A + \varepsilon_B) \pm \tfrac{1}{2}(\varepsilon_A - \varepsilon_B)\left(1 + \frac{4\beta^2}{(\varepsilon_A - \varepsilon_B)^2}\right)^{1/2}.$$

If the energy gap between A and B is large ($\varepsilon_A \ll \varepsilon_B$) and if the perturbation h' is small, then $x = 4\beta^2/(\varepsilon_A - \varepsilon_B)^2$ will be small and the square root in the expression above for ε can be approximated by $(1 + x)^{1/2} \sim (1 + \tfrac{1}{2}x)$ or

$$\varepsilon \sim \tfrac{1}{2}(\varepsilon_A + \varepsilon_B) \pm \tfrac{1}{2}(\varepsilon_A - \varepsilon_B)\left(1 + \frac{2\beta^2}{(\varepsilon_A - \varepsilon_B)^2}\right).$$

For the plus sign:

$$\varepsilon_1 = \varepsilon_A + \frac{\beta^2}{\varepsilon_A - \varepsilon_B} = \varepsilon_A + \Delta E,$$

and for the minus sign:

$$\varepsilon_2 = \varepsilon_B - \frac{\beta^2}{\varepsilon_A - \varepsilon_B} = \varepsilon_B - \Delta E,$$

where $\Delta E = \beta^2/(\varepsilon_A - \varepsilon_B)$. Now β^2 is positive, and since $\varepsilon_A \ll \varepsilon_B$, the quantity ΔE is negative. Thus, ε_1 is below ε_A and ε_2 is above ε_B (Fig. 3). That the amounts of energy raising and lowering are equal (ΔE) is a result of the neglect of overlap. Had overlap been included, the destabilization of ε_2 would have been greater than the stabilization of ε_1. The general rule is that

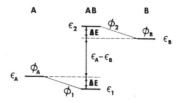

FIG. 3 Energy-level changes resulting from perturbation interactions between separated units A and B to form AB.

when two energy levels interact, the lower level is stabilized and the upper level is destabilized.

The perturbed orbitals ϕ_1 and ϕ_2 can be expressed as

$$\phi_1 = a_1\phi_A + b_1\phi_B \quad \text{and} \quad \phi_2 = a_2\phi_A + b_2\phi_B.$$

For each perturbed orbital, the ratio of coefficients b/a can be obtained by solving the homogeneous, linear equations [8], which for this problem reduce to

$$a(\varepsilon_A - \varepsilon) + b\beta = 0, \qquad [i]$$

$$a\beta + b(\varepsilon_B - \varepsilon) = 0. \qquad [ii]$$

For ϕ_1, $\varepsilon_1 = \varepsilon_A + \beta^2/(\varepsilon_A - \varepsilon_B)$. Substituting this value for ε into Eq. [i] gives

$$a_1[\varepsilon_A - \varepsilon_A - \beta^2/(\varepsilon_A - \varepsilon_B)] + b_1\beta = 0,$$

$$-a_1\beta/(\varepsilon_A - \varepsilon_B) + b_1 = 0,$$

$$b_1 = a_1\beta/(\varepsilon_A - \varepsilon_B).$$

For ϕ_2 substituting $\varepsilon_2 = \varepsilon_B - \beta^2/(\varepsilon_A - \varepsilon_B)$ into Eq. [ii] produces

$$a_2 = -b_2\beta/(\varepsilon_A - \varepsilon_B).$$

Therefore, the unnormalized perturbed orbitals are

$$\phi_1 = a_1[\phi_A + \phi_B\beta/(\varepsilon_A - \varepsilon_B)], \qquad \phi_2 = b_2[\phi_B - \phi_A\beta/(\varepsilon_A - \varepsilon_B)].$$

Now β is negative and small, and $\varepsilon_A \ll \varepsilon_B$. Therefore, the quantity $\beta/(\varepsilon_A - \varepsilon_B)$ is positive and small. The lower energy perturbed orbital ϕ_1 is mainly ϕ_A with a small amount of ϕ_B mixed in with the same phase. The higher energy orbital ϕ_2 is mostly ϕ_B with a small contribution of opposite phase from ϕ_A. The rule to remember is that when two orbitals interact, the lower energy orbital mixes into itself some of the higher energy one in a bonding way while the higher energy orbital mixes into itself some of the lower one in an antibonding way.

The heteronuclear two-electron ion HeH^+ illustrates the perturbation interaction rules. Call the two $1s$ AOs s_{He} and s_H. The energy of an electron in a 1s AO of a hydrogenlike ion is given by $E = -13.6Z^2$ eV, where Z is the nuclear charge. For He^+, $Z = 2$, and $\varepsilon(He^+) = -54.4$ eV. For H, $Z = 1$, $\varepsilon(H) = -13.6$ eV. The energy gap between unperturbed He^+ and H levels is large. The perturbed MOs are 1σ and 2σ (6). The ratio of coefficients in

$$1\sigma = a_1 s_{He} + b_1 s_H \qquad\qquad 2\sigma = a_2 s_{He} + b_2 s_H$$

6

1σ is $b_1/a_1 = \beta/(\varepsilon_A - \varepsilon_B)$. Suppose $\beta = \frac{1}{2}(\varepsilon_{He} + \varepsilon_H)S = -34S$ eV. For 1s AOs on He and H separated by their distance in HeH^+, the overlap is about 0.5. Therefore, $\beta = -17$ eV. Then $b_1/a_1 = \beta/(\varepsilon_A - \varepsilon_B) = 17/40.8 = 0.42$:

$$1\sigma = a_1(s_{He} + 0.42s_H).$$

The coefficient a_1 can then be chosen by normalization.

The ratio of coefficients a_2/b_2 in 2σ must be -0.42 from the perturbation treatment, but can also be determined by making 2σ orthogonal to 1σ. The relative sizes of the circles in **6** imply a larger coefficient for s_{He} than for s_H in 1σ and just the reverse in 2σ. However, the effective radius of s_{He} is smaller than that of s_H in both MOs because of the larger nuclear charge on He.

THE RELATIONSHIP AMONG ORBITAL ENERGIES, COEFFICIENTS, AND OVERLAPS

For a normalized MO ϕ_l the orbital energy ε_l is

$$\varepsilon_l = \int \phi_l H_{eff}\, \phi_l\, dv,$$

which, inserting the LCAO expansion for ϕ_l, Eq. [7], becomes (7)

$$\varepsilon_l = \sum_r c_{rl}^2 H_{rr} + \sum_{r \neq s} c_{rl}c_{sl} H_{rs}.$$

Assuming that H_{rs} is proportional to overlap, $H_{rs} = KS_{rs}$, the orbital energy ε_l becomes

$$\varepsilon_l = \sum_r c_{rl}^2 H_{rr} + K \sum_{r \neq s} c_{rl}c_{sl} S_{rs}. \qquad [10]$$

We will be concerned with orbital energy changes that result from changes in molecular geometry. Since molecular geometry enters extended Hückel calculations through the overlaps S_{rs}, energy changes arise in the second term of Eq. [10] suggesting

$$\Delta \varepsilon_l = K \sum_{r \neq s} c_{rl} c_{sl} \Delta S_{rs}. \qquad [11]$$

Recall that K is negative and overlaps are usually positive. If the AOs χ_r and χ_s have the same phase in ϕ_l (bonding combinations), the $c_{rl} c_{sl}$ will be positive and geometry changes that increase ΔS_{rs} will lower or stabilize the energy ε_l. Conversely, changes that increase overlap between AOs of opposite phase (antibonding combinations) destabilize the MO energy.

Equation [11] provides the basis for the application of qualitative MO theory. Orbital energy changes $\Delta \varepsilon_l$ will be large if the overlap changes ΔS_{rs} are large, or if the coefficients $c_{rl} c_{sl}$ are large, or if there are many reinforcing terms in the sum in Eq. [11]. Changes in AO overlaps can be estimated pictorially, the number of overlapping AO pairs can be counted and the phase relationships deduced, and even the relative importance of AO coefficients can be inferred. The higher energy MOs often have more nodes and the more nodes or phase changes produce larger AO coefficients that amplify the AO overlap changes. Therefore, it is often true that the highest occupied MO controls molecular properties because its energy changes are the largest.

OTHER RULES

Several other rules round out our initial survey of the qualitative MO tool kit.

a. Only valence electrons need be considered. Electrons in filled inner shells or atomic cores have such deep energies and small overlaps that their perturbations of the valence shell electrons are qualitatively negligible.

b. For the examples covered in the following chapters, the AO basis set consists of one $1s$ AO from each hydrogen or helium atom and valence s and p AOs from all other atoms. Compounds involving transition metals have been ignored in order to avoid including valence d AOs in the basis set. Although they follow the same qualitative rules as the s and p AOs, the d AOs increase the size of the basis set and, hence, the number of MOs that can be constructed and that must, therefore, be considered. Some properties of the compounds of the representative elements are often explained by invoking the participation of unoccupied d AOs in the bonding scheme. Qualitative MO theory can usually explain these same properties without assuming d AOs as part of the basis set. In these cases the omission of consideration of d AOs can be supported quantitatively by comparing the properties as calculated both with and without the d AOs in the basis set.

c. Trends in properties through a series of related molecules can often be correlated with trends in electronegativity differences between the constituent atoms. Mulliken has defined electronegativity as the average of the atomic ionization potential (IP) and the electron affinity (EA) (8). The IP is usually much larger than the EA and, therefore, the electronegativity roughly follows the IP. Now the IPs enter the extended Hückel method as the matrix elements H_{rr} or unperturbed AO energies. By relating IPs to electronegativities, we can hope to give MO explanations to electronegativity dependent properties.

d. It is well known that extended Hückel calculations break down when applied to highly ionic substances (9). Since qualitative MO theory is based on the approximations and formalism of the extended Hückel method, those substances with ionic or highly polar bonds must be excluded from consideration.

SUMMARY

The principal working rules of qualitative MO theory as applied in this book are now summarized, in some instances without their corresponding limitations or qualifications.

1. Consider valence electrons only.

2. Form completely delocalized MOs as linear combinations of valence s and p AOs.

3. MOs must be either symmetric or antisymmetric with respect to the symmetry operations of the molecule.

4. Compose MOs for structures of high symmetry and then produce orbitals for related but less symmetric structures by systematic distortions of the orbitals for higher symmetry.

5. The total energy is the sum of the orbital energies of individual valence electrons.

6. Molecules with similar molecular structures have qualitatively similar MOs and individual molecules in such classes differ primarily in the number of valence electrons that occupy the common MO system.

7. A change in the angular geometry of a molecule will produce a large change in the energy of a particular MO if the resulting changes in AO overlap are large. Even small overlap changes can produce large energy changes if the coefficients of the AOs involved are large or if there are many small overlap changes that can accumulate.

8. The AO coefficients are large in high energy MOs with many nodes or complicated nodal surfaces. The more AOs that occur in a MO of a particular nodal character, the smaller the coefficients of those AOs must be.

9. The more electronegative elements have lower energy AOs.

10. When two orbitals interact, the lower energy orbital is stabilized and the higher energy orbital is destabilized. An out-of-phase or antibonding interaction between two orbitals always raises the energy more than the corresponding in-phase or bonding interaction lowers the energy.

11. When two orbitals interact, the lower energy orbital mixes into itself the higher energy one in a bonding way while the higher energy orbital mixes into itself the lower energy one in an antibonding way.

12. If the two highest energy MOs of a given symmetry are composed of different kinds of AOs, then mix the two MOs to form hybrid orbitals.

13. Energies of orbitals of the same symmetry classification cannot cross each other. Instead, such orbitals mix and diverge.

REFERENCES

(1) H. F. Schaefer, III, "The Electronic Structure of Atoms and Molecules." Addison-Wesley, Reading, Massachusetts, 1972; L. C. Snyder and H. Basch, "Molecular Wave Functions and Properties." Wiley (Interscience), New York, 1972.

(2) R. S. Mulliken, C. A. Rieke, D. Orloff, and H. Orloff, *J. Chem. Phys.* **17**, 1248 (1949).

(3) J. Hinze and H. H. Jaffe, *J. Am. Chem. Soc.* **84**, 540 (1962); *J. Phys. Chem.* **67**, 1501 (1963).

(4) R. Hoffmann, *J. Chem. Phys.* **39**, 1397 (1963).

(5) R. S. Mulliken, *J. Chem. Phys.* **23**, 1833, 1841 (1955).

(6) L. Salem, *J. Am. Chem. Soc.*, **90**, 543 (1968); A. Imamura, *Mol. Phys.* **15**, 225 (1968); M. J. S. Dewar, "The Molecular Orbital Theory of Organic Chemistry." McGraw-Hill, New York, 1969; R. Hoffmann, *Accounts Chem. Res.* **4**, 1 (1971); R. F. Hudson, *Angew. Chem.* **85**, 63 (1973).

(7) J. P. Lowe, *J. Am. Chem. Soc.* **96**, 3759 (1974).

(8) R. S. Mulliken, *J. Chem. Phys.* **2**, 782 (1934).

(9) L. C. Allen and J. D. Russell, *J. Chem. Phys.* **46**, 1029 (1967).

2 The H_3 and H_4 activated complexes and related systems

The simplest polyatomic combinations for which MOs can be qualitatively constructed are those composed solely of hydrogen atoms, each of which contributes only a single $1s$ orbital to the AO basis set. In this chapter we consider some possible shapes for assemblages of three and four hydrogen atoms and then examine the elementary reaction processes in which H_3 and H_4 are assumed to occur as transition state complexes.

MOs AND SHAPES OF H_3

Three $1s$ AOs, one from each hydrogen atom, combine to form three MOs for the H_3 complex. Figure 1 shows AO composition diagrams for each MO and compares relative MO energies for H_3 in linear symmetric $(D_{\infty h})$ and equilateral triangular (D_{3h}) geometries. Assume all distances between connected hydrogens are equal and constant in both structures. For linear H_3, the MO energies increase with the number of nodal surfaces: $1\sigma_g$ (nodeless) $< 1\sigma_u$ (one node) $< 2\sigma_g$ (two nodes). For triangular H_3, the lowest energy MO is a_1', the nodeless combination of three in-phase overlapping $1s$ AOs. Symmetry allows a doubly degenerate MO pair e' at higher energy. (See Chapter 1.) Differences in AO overlaps relate the energies of MOs between the two extreme geometries. The end hydrogen AOs in $1\sigma_g(D_{\infty h})$ do not overlap each other as well as they do in $a_1'(D_{3h})$, giving a_1' a lower energy. Energy differences between the conformations of the nodeless MO are never large, one reason being that the coefficients are generally

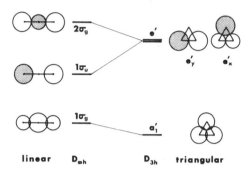

linear $D_{\infty h}$ D_{3h} triangular

FIG. 1   MO correlations between linear and triangular structures for H_3 showing qualitative AO compositions of each MO. (After Ref. 15.)

small since they all have the same sign. There is a much larger energy difference between $1\sigma_u(D_{\infty h})$ and $e'(D_{3h})$. The energy rises along $1\sigma_u$-e_y' as bending pushes $1s$ AOs on the end hydrogens towards each other from opposite sides of the nodal surface and into greater out-of-phase overlap with each other. The energy decreases from $2\sigma_g(D_{\infty h})$ to $e_x'(D_{3h})$ as end AOs come together in-phase. Remember that the coefficients of e_y' are larger than those for the same hydrogens in either e_x' or a_1'. Since energy changes are proportional to overlap changes weighted by products of coefficients, the $e'(D_{3h})$ levels will be above the average of $1\sigma_u$ and $2\sigma_g(D_{\infty h})$. This is another aspect of the rule that increasing out-of-phase interactions raise the energy more than comparable in-phase interactions lower it. The result is that $1\sigma_u$-e' controls molecular shapes when it contains even one electron.

The ion $H_3{}^+$ has been observed experimentally and it is known to be rather strongly bound. With only two electrons, both in the $1\sigma_g$-a_1' MO system, $H_3{}^+$ has minimum energy when its shape is that of an equilateral triangle, the arrangement in which each $1s$ AO can achieve maximum overlap with the other two AOs. A third electron in $1\sigma_u$-e_y' gives H_3 linear shape by minimizing the out-of-phase overlaps between the two end $1s$ AOs. All three-, four-, and five-electron systems should be linear, while one- and two-electron systems should be triangular. These rules predict shapes for H_3^{2+}, $H_3{}^+$, H_3, $H_3{}^-$, and $He_3{}^+$ that agree with the results of the most rigorous ab initio calculations, our only reliable source of structures for these simple systems (1–5).

THE H₄ COMPLEX

Unlike H_3, which can be only either linear or bent, there are many possible shapes for H_4. Two simple structures for H_4 are the square (D_{4h}) and the linear $(D_{\infty h})$ shapes. Figure 2 correlates the MO energies for these

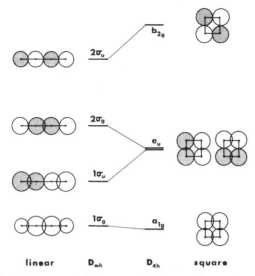

FIG. 2   Qualitative MO correlation diagram for linear and square H₄. (After Ref. 15.)

two structures. Assume all distances between connected H atoms are equal and constant. The linear MOs with higher energy have more nodal surfaces. By bending up the terminal hydrogens of the linear structure through cis geometry to the square, the AO compositions of the MOs of the square become readily apparent. It is easy to see that linear MOs $1\sigma_u$ and $2\sigma_g(D_{\infty h})$ become exactly degenerate as the $e_u(D_{4h})$ orbitals in the square shape. The energy decreases slightly along the nodeless MO system $1\sigma_g(D_{\infty h})$-$a_{1g}(D_{4h})$ as terminal AOs move together to overlap in-phase. The energy of $1\sigma_u(D_{\infty h})$-$e_u(D_{4h})$ rises more sharply because comparable out-of-phase interactions are larger. Therefore the four-electron H₄ complex has lower energy as a linear structure than as a square.

Imagine forming a tetrahedron (T_d) by flexing a square (D_{4h}), as shown in **1**. Figure 3 compares MO energies for square and tetrahedral H₄. Again,

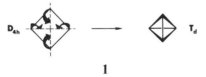

1

assume constant and equal distances between all connected hydrogens. The $a_1(T_d)$ MO must have the lowest energy of all possible combinations of four $1s$ AOs since each AO overlaps in-phase with three others at bonding distances. Indeed, we can immediately conclude that the two-electron ion H_4^{2+} should be tetrahedral, as ab initio calculations indicate that it is (6).

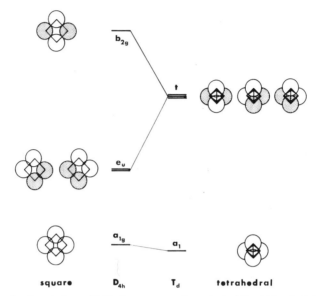

FIG. 3   Correlations of MOs for square and tetrahedral H_4. (After Ref. 15.)

The accompanying tabulation counts numbers of in-phase and out-of-phase overlaps between one $1s$ AO and its nearest neighbors in some square and tetrahedral MOs, providing a measure of relative energies of those MOs.

	In-phase overlaps	Out-of-phase overlaps
$b_{2g}(D_{4h})$	0	2
$t_2(T_d)$	1	2
$e_u(D_{4h})$	1	1

The triply degenerate $t_2(T_d)$ MOs lie in energy below b_{2g} but above $e_u(D_{4h})$. Since $t_2(T_d)$ is higher than $e_u(D_{4h})$ the tetrahedral structure of H_4 is less stable than the square shape.

The four-electron H_4 complex must have two electrons in a lower energy, nodeless MO and two electrons in a higher energy orbital with one nodal surface. The out-of-phase interactions that arise because of the node in the higher occupied orbital determine the relative energies of possible H_4 shapes. From this argument one can see that the tetrahedral structure must have the highest energy of all possible H_4 structures because $t_2(T_d)$ has the maximum possible number of out-of-phase interactions for a MO with a single node. Furthermore, the lowest energy H_4 structure must be linear because $1\sigma_u$ has only one pair of out-of-phase AOs at connecting distances and two pairs of in-phase AOs. Therefore, we can establish a clear order of H_4 structural stabilities:

linear > square > tetrahedral.

Some excellent ab initio calculations verify these qualitatively deduced stability conclusions for H_4 shapes (7).

REACTION MECHANISMS

The hydrogen-deuterium exchange reaction $H_2 + D_2 \rightarrow 2HD$ has been studied experimentally both in high-temperature flow reactors (8) and in shock tubes (9). In the flow reactors it is known that H_2 and D_2 dissociate into atoms at the hot walls of the reaction vessel and these heterogeneously produced atoms then react to form HD product through the homogeneous gas phase propagation steps:

$$H\cdot + D_2 \longrightarrow HD + D\cdot$$
$$D\cdot + H_2 \longrightarrow HD + H\cdot \qquad [1]$$

The experimental activation energy is about 10 kcal/mole.

In the shock tube experiments, the gases are heated to high temperature for only a very short time while the walls of the reaction vessel remain cool, eliminating the possibility of heterogeneous dissociations of H_2 and D_2 and the free radical chain reactions [1] while allowing the slower homogeneous bimolecular reaction [2] to occur:

$$H_2 + D_2 \longrightarrow 2HD \qquad [2]$$

The experimental activation energy is 42 kcal/mole.

Molecular quantum mechanics has solved the problem of the chain propagation mechanism [1]. The reaction goes through a linear H_3 (or HD_2 or DH_2) complex and the calculated activation energy E_a agrees quantitatively with experimental results (Fig. 4).

Figure 5 correlates occupied orbitals between reactants, linear transition state, and products for the reaction $H + H_2 \rightarrow H_2 + H$. For an end-on collision of a hydrogen atom with a hydrogen molecule, all the MOs have

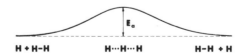

FIG. 4   Schematic energy variation for the hydrogen exchange reaction proceeding through a linear, symmetric transition state.

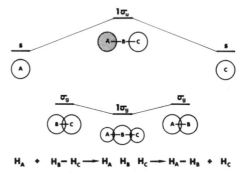

H_A + H_B– H_C ⟶ H_A H_B H_C ⟶ H_A– H_B + H_C

FIG. 5   Correlations of occupied orbitals for reactants H + H₂ through the linear H₃ transition state to form the exchanged products H₂ + H.

cylindrical or σ symmetry throughout the process. Therefore, the MO labels in Fig. 5 are those for the isolated species and not for the overall $C_{\infty v}$ process. Electrons flow smoothly from occupied orbitals of the reactants, through those for the linear H₃ transition state, to the lowest available orbitals of the appropriate products. The activation energy for the reaction must arise from the hump in the energy of the s-$1\sigma_u$-s MO system that carries one electron from an s AO of an isolated hydrogen atom through the nonbonding $1\sigma_u$ MO of the H₃ complex to the s AO of the opposite H atom in the products.

The reaction He + H_2^+ → HeH$^+$ + H has been studied experimentally (10). It must also proceed through a linear, three-electron transition state HeHH$^+$ (11). Figure 6 is a correlation diagram for the reaction. The nodal properties of the HeHH$^+$ MOs are similar to those of the higher symmetry H₃ complex. Comparing energies of reactant orbitals, the He or He$^+$ AO s_{He} should be lower than the σ_g MO of H₂ or H_2^+. The low energy s_{He} should be only slightly perturbed by σ_g or s in intermediate and product. The 1σ MO of HeH$^+$ should be below the nonbonding s_H AO of an isolated hydrogen

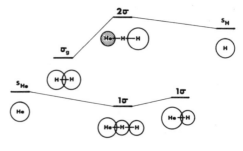

He + H – H ⟶ He ⋯ H ⋯ H ⟶ He – H + H

FIG. 6   Correlations of occupied orbitals for the reaction He + H_2^+ → HeH$^+$ + H.

atom. Again, electrons flow smoothly from reactants to the proper products. The electron pair in s_{He} of the reacting He atom passes through $1\sigma\,(HeH_2{}^+)$ to $1\sigma\,(HeH^+)$. The unpaired electron in $1\sigma\,(H_2{}^+)$ moves through $2\sigma\,(HeH_2{}^+)$ to the s_H AO of the neutral hydrogen atom product.

The reaction between He^+ and H_2 to produce HeH^+ and H has not been observed (12) and Fig. 6 explains why. With reactant orbital s_{He} singly occupied and two electrons in $\sigma_g\,(H_2)$, the allowed products turn out to be the unlikely ions H^- and HeH^{2+}.

Unlike the H_3 problem, molecular quantum mechanics has thus far not been able to confirm the experimental activation energy or elucidate the structure of the H_4 complex. Although it has the lowest energy, linear H_4 is not a possible transition state complex because it cannot lead to isotope exchange without dissociation of hydrogen atoms, which would be very costly in energy. Neither the square nor the tetrahedron are satisfactory shapes either, as the following symmetry arguments explain.

Figure 7 correlates MOs from reactants $H_2 + D_2$ through a square transition state to produce 2HD. The tie-lines connecting energy levels in Fig. 7 have been labeled to indicate the rectangular D_{2h} symmetry of the overall process. A pair of electrons starting out in the in-phase combination of two separated $\sigma_g\,(H_2, D_2)$ MOs moves through a_g to $a_{1g}(H_4)$ to a_g and then back to the separated but in-phase σ_g MOs of HD product. The other occupied MO results from the out-of-phase combination of two σ_g reactant MOs and becomes b_{1u}, which rises through e_y'-b_{1u} to the in-phase combination of two antibonding σ_u MOs of products. At the same time, empty σ_u

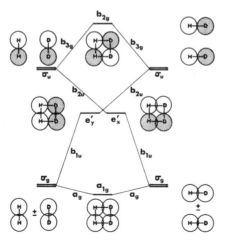

FIG. 7   Orbital correlations for the reaction $H_2 + D_2 \rightarrow$ 2HD involving the four-center, square transition state. This process is symmetry forbidden because of the crossing of filled b_{1u} and empty b_{2u} orbitals in the square transition state.

MOs of reactants combine in-phase and fall in energy, passing through b_{2u}-e_x'-b_{2u} to produce the out-of-phase combination of σ_g MOs of product. Because of the crossing of filled (b_{1u}) and empty (b_{2u}) orbitals in the square transition state, this mechanism violates the principle of conservation of orbital symmetry of Woodward and Hoffmann (13,14). Experience shows that when filled and empty MOs cross in the transition state, the activation barrier is usually too high to allow the reaction to proceed by thermal activation.

The symmetry forbidden nature of the square, four-center mechanism can be described in a more fundamental way using configuration interaction terminology. On the reactant side of the square transition state, the electronic configuration is best represented by $a_g^2 b_{1u}^2$ while that on the products side is $a_g^2 b_{2u}^2$. At the transition state itself, neither configuration alone provides a satisfactory description of the lowest electronic state. A better representation of the electronic structure of geometries intermediate between reactants and products would be a linear combination of the two configurations:

$$\Phi_{CI} = c_1 a_g^2 b_{1u}^2 + c_2 a_g^2 b_{2u}^2.$$

The mixing of configurations or configuration interaction (CI) gives a lower energy than does either configuration independently, but it is likely that the energy in the square structure will still be higher than energies at the extremes of separated reactants and products and that there will be a substantial activation barrier separating reactants and products (Fig. 8).

MO considerations similar to those for square H_4 lead to the conclusion that tetrahedral H_4 also presents a symmetry forbidden pathway to isotope exchange (2).

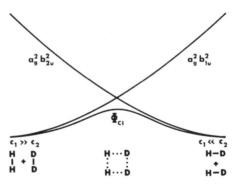

FIG. 8   The activation barrier for the square planar transition state as a result of configuration interaction.

2

Ab initio calculations, many with extensive configuration interaction, have been carried out for H_4 structures, including linear ($D_{\infty h}$), square (D_{4h}), tetrahedral (T_d), centered equilateral triangular (D_{3h}), rhombic (D_{2h}), rectangular (D_{2h}), trapezoidal, T and Y shapes (C_{2v}) (7). The results of these calculations show that linear H_4 has an energy 40 to 45 kcal/mole above that of separated $2H_2$, an energy comparable to the experimental activation energy. The square has an energy that is above $2H_2$ by perhaps 120–140 kcal/mole, even higher than the 109 kcal/mole required to dissociate H_2 into 2H and initiate the chain reaction [1]. The tetrahedron apparently does have the highest energy of all the proposed H_4 structures with the energies of the other conformations falling somewhere between those of the square and the tetrahedron. Since all but the linear H_4 structure have energies above that of $H_2 + 2H$, there seems to be no energetically available path through which a four-center mechanism could pass. Furthermore, the proposed H_4 structures are at least 60–80 kcal/mole higher than the activation energy observed in the shock tube experiments. The descrepancies are are much larger than the estimated errors in the ab initio calculations. It may be that the shock tube experiments are not actually monitoring a bimolecular exchange reaction after all. The $H_2 + D_2$ mechanism has yet to be resolved.

REFERENCES

(1) H_3^{2+}: H. Conroy, *J. Chem. Phys.* **51**, 3979 (1969).

(2) H_3^+: R. E. Christoffersen, S. Hagstrom, and F. Prosser, *J. Chem. Phys.* **40**, 236 (1964); R. E. Christoffersen, *J. Chem. Phys.* **41**, 960 (1964); L. Salmon and R. D. Poshusta, *J. Chem. Phys.* **59**, 3497 (1973).

(3) H_3: H. Conroy and B. L. Bruner, *J. Chem. Phys.* **42**, 4047 (1965); H. H. Michels and F. E. Harris, *J. Chem. Phys.* **48**, 2371 (1968); I. Shavitt, R. M. Stevens, F. L. Minn, and M. Karplus, *J. Chem. Phys.* **48**, 2700 (1968); C. Edmiston and M. Krauss, *J. Chem. Phys.* **49**, 192 (1968); B. Liu, *J. Chem. Phys.* **58**, 1925 (1973).

(4) H_3^-: A. Macias, *J. Chem. Phys.* **48**, 3464 (1968); **49**, 2198 (1968).

(5) He_3^+: R. D. Poshusta and D. F. Zetik, *J. Chem. Phys.* **48**, 2826 (1968); C. Vauge and J. L. Whitten, *Chem. Phys. Lett.* **13**, 541 (1972).

(6) H_2^{2+}: A. A. Frost, *J. Chem. Phys.* **47**, 3714 (1967).

(7) H_4: H. Conroy and G. Malli, *J. Chem. Phys.* **50**, 5049 (1969); M. Rubenstein and I. Shavitt, *J. Chem. Phys.* **51**, 2014 (1969); C. W. Wilson, Jr. and W. A. Goddard III, *J. Chem. Phys.* **51**, 716 (1969); **56**, 5913 (1972); C. F. Bender and H. F. Schaefer III, *J. Chem. Phys.* **57**, 217 (1972); M. E. Schwarz and L. J. Schaad, *J. Chem. Phys.* **43**, 4709 (1968);

B. Freihaut and L. M. Raff, *J. Chem. Phys.* **58**, 1202 (1973); D. M. Silver and R. M. Stevens, *J. Chem. Phys.* **59**, 3378 (1973).

(8) A. Farkas and L. Farkas, *Proc. R. Soc. London* **A152**, 124 (1935); M. van Meersche, *Bull. Soc. Chim. Belg.* **60**, 99 (1951); G. Boato, G. Careri, A. Cimino, E. Molinari, and G. G. Volpi, *J. Chem. Phys.* **24**, 783 (1956).

(9) S. H. Bauer and E. Ossa, *J. Chem. Phys.* **45**, 434 (1966); A. Burcat and A. Lifshitz, *J. Chem. Phys.* **47**, 3079 (1967); R. D. Kern and G. G. Nika, *J. Phys. Chem.* **75**, 1615, 2541 (1971).

(10) R. H. Neynaber and G. D. Magnuson, *J. Chem. Phys.*, **59**, 825 (1973).

(11) C. Edmiston, J. Doolittle, K. Murphy, K. T. Tang, and W. Wilson, *J. Chem. Phys.* **52**, 3419 (1970); P. J. Brown and E. F. Hayes, *J. Chem. Phys.* **55**, 922 (1971).

(12) J. J. Leventhal, *J. Chem. Phys.* **54**, 3279 (1971).

(13) R. Hoffmann, *J. Chem. Phys.* **49**, 3739 (1968).

(14) R. B. Woodward and R. Hoffmann, "The Conservation of Orbital Symmetry." Academic Press, New York, 1970.

(15) B. M. Gimarc, *J. Chem. Phys.* **53**, 1623 (1970).

3 The shapes of AH₂, AH₃, and AH₄ molecules

The classes of molecules and ions with general formulas AH_2, AH_3, and AH_4, where A is an atom other than hydrogen, contain some of the most important examples for qualitative explanations of chemical valence. The high symmetries of these molecules and the small numbers of AOs in the basis sets make the formation of qualitative MOs particularly easy. The principal concern of this chapter is the rough features of molecular geometry in these classes, but other properties such as trends in valence angles, relative energies of electronic states, and barriers to inversion can also be understood using the same theoretical models.

AH_2

Table I lists the AH_2 molecules and ions for which experimental ([1–19]) and theoretical ([20–35]) structural data are available. With only three atoms these molecules can be only either linear or bent in shape. Molecules with three or four valence electrons are linear; those with one or two or five through eight electrons are bent.

From the s and three p valence AOs on the central atom A and the $1s$ orbitals from the two hydrogens, a total of six valence MOs for AH_2 can be formed. These are shown in Fig. 1. The MOs for linear ($D_{\infty h}$) AH_2 are particularly easy to describe. The lowest energy valence orbital, $1\sigma_g$, is composed of the s AO on A plus in-phase overlapping $1s$ AOs from the

TABLE I

The shapes of AH_2 molecules[a]

AH_2	Number of valence electrons	Shape
$LiH_2{}^+$	2 }	bent
LiH_2, $BeH_2{}^+$	3 ⎫	linear
BeH_2, $BH_2{}^+$	4 ⎭	
BH_2, AlH_2, $CH_2{}^+$	5 ⎫	
CH_2, SiH_2, GeH_2, $BH_2{}^-$, $NH_2{}^+$	6 ⎪	
NH_2, PH_2, AsH_2, $CH_2{}^-$, $SiH_2{}^-$, $OH_2{}^+$, $SH_2{}^+$	7 ⎬	bent
H_2O, H_2S, H_2Se, H_2Te, $NH_2{}^-$, $PH_2{}^-$, H_2F^+	8 ⎭	

[a] For experimental and calculated structural data, see Refs. 1–35.

hydrogens. The p_z AO of A points directly at the 1s orbitals on the hydrogens that overlap in-phase with p_z to form the $1\sigma_u$ MO that has lower energy than p_z alone. In the two $1\pi_u$ orbitals, the hydrogens are located on the nodal surfaces of the p_x and p_y orbitals and hence the hydrogen 1s AOs cannot enter either $1\pi_{ux}$ or $1\pi_{uy}$. These two MOs are pure p_x or p_y and they are degenerate in energy. The antibonding $2\sigma_g$ orbital is the out-of-phase combination of central atom s and hydrogen 1s AOs. Because the overlaps are

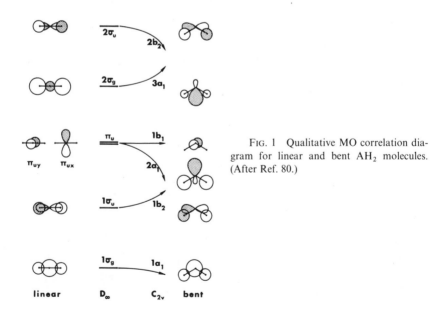

FIG. 1 Qualitative MO correlation diagram for linear and bent AH_2 molecules. (After Ref. 80.)

larger, the out-of-phase interaction of p_z with two hydrogen AOs in $2\sigma_u$ should be larger than that among the s orbitals in $2\sigma_g$, giving $2\sigma_u$ the highest energy of the six MOs of linear geometry. In Fig. 1 and all subsequent MO diagrams in this book, the numbering scheme for MOs covers valence and higher energy orbitals only. For example, in H$_2$S the inner shell or core MOs of σ_g symmetry composed primarily of the 1s and 2s AOs of sulfur were ignored when labels were assigned to the occupied valence and empty MOs 1σ_g and 2σ_g.

Figure 1 correlates orbitals and energies for linear ($D_{\infty h}$) and bent (C_{2v}) AH$_2$. On bending, the two hydrogen 1s AOs in 1σ_g move closer together in 1a_1 and therefore into better overlap with each other, giving 1a_1 (C_{2v}) lower energy that the related 1σ_g ($D_{\infty h}$). If the molecule contains only one or two valence electrons, these will find lower energy in bent geometry in 1a_1. Ions such as H$_3^+$ and LiH$_2^+$ are indeed bent. In H$_2$O, for example, the 1σ_g-1a_1 MO is mainly the oxygen 2s AO slightly perturbed by the hydrogen 1s AOs. This is so because the oxygen 2s is deep in energy (-36 eV) compared to the hydrogen 1s AOs (-13.6 eV). Therefore, the hydrogen AO coefficients in 1σ_g-1a_1 are small making the energy differences between 1σ_g and 1a_1 small. When higher energy MOs are occupied, this energy difference is negligible. Similarly, perturbation interactions make the 1s AOs the major components of the antibonding 2σ_g MO, with a smaller contribution from the central atom s AO. However, a different perturbation situation occurs for LiH$_2^+$. The lithium 2s AO is high (-5.4 eV) compared to the bonding MO for H$_2$ (-16 to -20 eV). In LiH$_2^+$, the 1a_1 MO is principally σ_g of H$_2$ slightly perturbed by the 2s AO of Li$^+$ (**1**). This description is

1

supported by ab initio calculations which show that LiH$_2^+$ has an isosceles triangular structure in which H$_2$, with the H-H bond distance practically unchanged from that in free H$_2$, weakly interacts with a Li$^+$ ion at a distance of 1.8 to 2.0 Å (*20*).

The energy of the 1σ_u ($D_{\infty h}$) increases on bending because the hydrogen 1s AOs are pulled out of overlap with the lobes of the p_z AO on the central atom and pushed towards each other across the nodal surface of the orbital 1b_2(C_{2v}) of bent geometry. The 1σ_u-1b_2 MO controls the shapes of three- and four-electron AH$_2$ molecules such as BeH$_2$ and BH$_2^+$, which are linear because the electrons are at lower energy in the linear 1σ_u orbital. But another possibility exists which we discuss later.

When $1\pi_{uy}$ is bent, the hydrogens move on the nodal surface of $1b_1$. No overlap change is possible and therefore $1\pi_{uy}$ $(D_{\infty h})$ and $1b_1(C_{2v})$ have the same AO composition and the same energy. The $1\pi_{uy}$-$1b_1$ MO system has no geometry preference and no direct influence on molecular shapes. As $1\pi_{ux}$ is bent, the hydrogens move off the nodal surface of p_x and in-phase overlapping of the hydrogen $1s$ orbitals with the bottom lobe of p_x lowers the energy of the orbital $2a_1$ thus formed (2). The shapes of molecules with

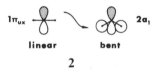

2

five through eight electrons are bent because $2a_1$ falls rapidly in energy as the valence angle HAH decreases. In molecules such as BH_2 and AlH_2, the $2a_1$ orbital contains only one electron. Since BH_2 and AlH_2 are both bent, the energy of the singly occupied $2a_1$ orbital must fall at least twice as rapidly as the energy of the doubly occupied $1b_2$ rises below it. That this might be so can be seen from the following overlap argument. The overlap between a p orbital and a hydrogen $1s$ orbital is proportional to the cosine of the angle ϕ between the axis of the p orbital and the A-H bond axis. Figure 2 traces the first quarter period of the cosine function. The position of maximum overlap has the hydrogen $1s$ AO resting squarely on the p orbital axis ($\phi = 0$). With the hydrogen tilted 30° away from the p-orbital axis, the $1s$, p overlap is still 87% of maximum and at $\phi = 60°$ the overlap is 50% of maximum. At $\phi = 90°$, the $1s$ AO is on the p-orbital nodal surface and the

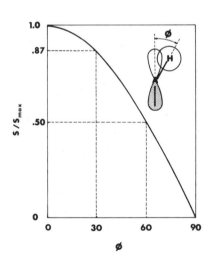

Fig. 2   Overlap dependence on the angle ϕ between the A-H bond and the p-orbital axis. (From Ref. 80.)

overlap is zero. If the size of the energy change is related to the amount of overlap change, then tilting a hydrogen a few degrees away from the p-orbital axis should produce a relatively small energy increase. Moving the hydrogen a few degrees off the p-orbital nodal surface should lead to a significant energy lowering. For example, imagine decreasing the angle HAH from 180° (linear) to 120° (bent). Each of two $1s$, p_z overlaps in $1b_2$ decreases from 100 to 87% of maximum, producing a modest rise in energy, while each of two $1s$, p_x overlaps in $2a_1$ increases from 0 to 50% of maximum for a substantial energy lowering or stabilization. Therefore, even a single electron in $2a_1$ can force molecules such as BH$_2$ and AlH$_2$ to be bent.

In its ground state BH$_2$ has a valence angle of 131°. The linear excited state of BH$_2$ is also known for which the electron configuration is $(1a_1)^2(1b_2)^2(1b_1)^1$. In this case an electron has been removed from the MO $2a_1$ that holds ground state BH$_2$ bent; the excited electron occupies $1b_1$, which has no geometry preference and the excited molecule is therefore linear because of the electron pair in $1\sigma_u$-$1b_2$. The same argument explains the structure of AlH$_2$, which is bent (119°) in its ground state but linear in its first excited state.

The arguments that led to a linear structure prediction for the three-electron species may be inappropriate or incomplete for LiH$_2$ and BeH$_2{}^+$. Both Li and Be have $2s$ AO energies that are higher than that of the hydrogen $1s$ AO and well above that of the σ_g MO of H$_2$. Recall that the perturbation interaction of σ_g (H$_2$) with $2s$ (Li$^+$) produces an acute valence angle in LiH$_2{}^+$. For small valence angles, $2a_1$ may cross and fall below $1b_2$, giving rise to two stable electron configurations: an acutely bent $(1a_1)^2(2a_1)^1$ and a linear $(1\sigma_g)^2(1\sigma_u)^1 = (1a_1)^2(1b_2)^1$. Semiempirical calculations for LiH$_2$ (21) predict that both linear and acutely bent ($\sim$20°) states of LiH$_2$ are stable, with the bent form about 40 kcal/mole more stable than the linear one. The results of ab initio SCF MO and valence bond calculations (22) also predict BeH$_2{}^+$ to have two different states or energy minima: an acutely bent form ($\sim$20°) and a higher energy structure with a wider ($\sim$90°) valence angle.

Simply assigning electrons to the lowest available MO energy levels for AH$_2$ leads one to the conclusion that for six-valence electron molecules the $(2a_1)^2$ singlet configuration should have lower energy than the $(2a_1)^1(1b_1)^1$ triplet. Both experiment (6) and ab initio SCF MO calculations (27) show that this conclusion is wrong for CH$_2$ and apparently for NH$_2{}^+$ as well (30). The ground states of these species are, in fact, triplets $(2a_1)^1(1b_1)^1$. The valence angles for these ground-state triplets are 136° (CH$_2$) and 140°–150° (NH$_2{}^+$), rather wide angles, comparble to that for the BH$_2$ ground state and consistent with the notion that $2a_1$ is singly occupied. Each of these species has a singlet state at higher energy with $2a_1$ doubly occupied and hence a sharper valence angle. For the $(2a_1)^2$ singlets, the angles are $105°$ (CH_2) and $115°$–$120°$ (NH_2^+). The singlets of still higher energy, the $(2a_1)^1(1b_1)^1$ configuration, have wide valence angles similar to those of the ground-state triplets: $140°$–$150°$ for CH_2 and around $150°$ for NH_2^+. The ground state of SiH_2 is a singlet $(2a_1)^2$ and has a valence angle of $92.8°$. Higher in energy are the singlet and triplet states with the configuration $(2a_1)^1(1b_1)^1$ and the wider valence angle of $123°$. Although the sums of orbital energies fail to give the proper energy order of states of CH_2 and NH_2^+, the qualitative model does represent the correct trends in bond angles for those states.

Qualitative MO theory does not adequately include electron-electron repulsions and it accounts for none of the differences in repulsions among the same number of electrons in different electron configurations. Clearly, the paired electron configuration $(2a_1)^2$ should produce higher repulsions than the unpaired configuration $(2a_1)^1(1b_1)^1$. These differences can be included semiempirically in extended Hückel calculations by adding an estimate of electron interactions to the sum of one-electron orbital energies for the singlet $(2a_1)^2$ configuration (36). In a purely qualitative model an electron interaction term may serve as a rationalization for the lower energy of CH_2 and NH_2^+ triplet states, but it provides no basis for predictions.

The ground-state electron configuration of the seven-valence electron species is $(2a_1)^2(1b_1)^1$ and valence angles of $103°$ for NH_2 and $99°$ for CH_2^- are consistent with double occupancy of $2a_1$. Excitation to the configuration $(2a_1)^1(1b_1)^2$ moves one electron from the orbital that holds NH_2 bent to an orbital with no geometry preference. A single electron in $2a_1$ is enough to keep excited NH_2 bent but at the wider valence angle of $144°$. PH_2 shows a similar increase of valence angle from the ground state ($91.70°$) to excited state ($123.2°$).

AH_2 molecules with eight-valence electrons, such as water, are bent due to double occupancy of the $2a_1$ orbital.

The high symmetry of linear geometry restricts the AO compositions of the linear MOs to those represented in Fig. 1. In many cases the AO compositions of the MOs for bent geometry are qualitatively the same as those for the related orbitals of linear shape and the energy differences are easily visualized by inspecting the AO overlap changes. The π_u-$1b_1$ and $1\sigma_u$-$1b_2$ orbitals in Fig. 1 are examples of such cases. Of course, $1s$ AOs, not allowed in $1\pi_{ux}$ by symmetry, had to be added to $2a_1$. There are still other examples in which the bent MOs produced by simple deformation of the linear MOs do not have the complete AO composition that symmetry allows. Consider, for example, the three MOs of a_1 symmetry (3). The symbols in parentheses are the classifications of the related linear MOs. Although symmetry

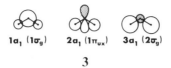

1a$_1$ (1σ_g)   2a$_1$ (1π_{ux})   3a$_1$ (2σ_g)

3

eliminates the p_y AO of the central atom from any participation in 1σ_g and 2σ_g ($D_{\infty h}$), the p_y orbital does have the proper symmetry to enter the related bent orbitals 1a$_1$ and 3a$_1$(C_{2v}). Similarly, the central atom s AO cannot enter 1π_u ($D_{\infty h}$) but it could add to 2a$_1$(C_{2v}). More realistic representations of the bent MOs can be obtained by mixing together the MOs produced directly by deformation of the linear MOs. Of course, the before mixing state of MOs is hypothetical; MOs that come from a MO calculation would appear already mixed. In general, the mixing becomes more important the higher the energy of the orbitals. All of the examples in this book are adequately covered by the following two rules for mixing of MOs of the same symmetry: (i) Mix only the highest energy pair of MOs (whether occupied or not), unless a pair of lower energy orbitals turn out to have particularly close energies, then, (ii) mix any pair of MOs that happen to be close in energy. Applying these rules to the AH$_2$ MOs of a_1 symmetry, we accept the unmixed picture of 1a$_1$ as a reasonable representation, but we must mix the pictures above for 2a$_1$ and 3a$_1$, the highest energy pair of a_1 symmetry. Figure 3 shows the process. The principal effect on AO composition that results from 2a$_1$-3a$_1$ mixing is the combination of central atom s and p AOs to form something that resembles an sp hybrid orbital on A. In 2a$_1$ the hybrid points away from the vertex of the valence angle, while in 3a$_1$ it points between the hydrogen 1s AOs and with opposite phase. The energy effect of MO mixing is to stabilize further the lower energy MO, in this case 2a$_1$,

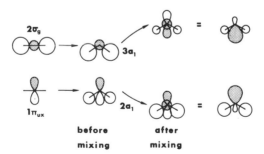

FIG. 3   The 2σ_g and 1π_{ux} orbitals of linear geometry become the highest energy pair of a_1 symmetry in bent geometry. Mixing of 2a$_1$ and 3a$_1$ corrects for AOs not allowed in the related linear orbitals. An important result is the hybrid form of 2a$_1$.

and to destabilize or raise the energy of the higher energy MO, $3a_1$. That $3a_1$ should rise in energy as the valence angle is decreased is readily apparent from the after-mixing AO composition of that MO. The unmixed AO composition of $3a_1$ would lead one to expect bending to lower the MO energy because of increased overlap between the hydrogen $1s$ AOs.

The valence bond picture of water shows two equivalent lone pair orbitals on the oxygen atom. In the MO model the lone pairs of electrons occupy the nonequivalent $2a_1$ and $1b_1$ MOs, a description that is in accord with the photoelectron spectrum of water (37).

What about the mixing of $1b_2$ and $2b_2$, another pair of MOs of the same symmetry? These two orbitals have the same AO composition and orbital mixing could only change the relative weights of hydrogen $1s$ and central atom s AOs in each MO. These weights are usually more easily introduced by perturbation arguments, such as those we used for $1\sigma_g$ and $2\sigma_g$. Generally, the mixing of orbitals of the same symmetry need be done only for MOs that are composed of different kinds of AOs.

Orbital mixing considerations provide explanations of the variation in valence angles through bent, isoelectronic series and trends in the energy difference between singlet and triplet states of the six-electron series. The extent of mixing between MOs $2a_1$ and $3a_1$ is determined by the energy gap between the unmixable linear MOs to which they are related: $1\pi_{ux}$ and $2\sigma_g$. The closer together $1\pi_{ux}$ and $2\sigma_g$, the more mixing between $2a_1$ and $3a_1$. Compare the AO compositions of $1\pi_{ux}$ and $2\sigma_g$ for a molecule such as water that has an electronegative central atom, that is, a central atom with deep lying valence s and p AOs (4). The $1\pi_{ux}$ orbital is a pure p AO on A. The

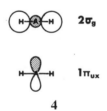

4

energy of $1\pi_{ux}$ will be higher the less electronegative the central atom A. The MO $2\sigma_g$ is the antibonding combination of the same AOs that constitute $1\sigma_g$. Perturbation interactions make $2\sigma_g$ largely the $1s$ AOs on the hydrogens with a smaller contribution from the s AO of A. The energy of $2\sigma_g$ will be lower for longer A-H bonds that decrease the antibonding interactions between hydrogen $1s$ and central atom s AOs. The A-H bond distance is determined by the in-phase AO overlaps in the occupied MOs at lower energy. Remember that for molecules with eight or fewer valence electrons $2\sigma_g$ is unoccupied. The larger the effective radius of the central atom AOs,

TABLE II

Typical A-H bond distances

AH$_2$	Configuration	R(A-H) (Å)
BH$_2$	$(2a_1)^1$	1.181
CH$_2$	$(2a_1)^2$	1.11
NH$_2$	$(2a_1)^2(1b_1)^1$	1.024
OH$_2$	$(2a_1)^2(1b_1)^2$	0.958
SH$_2$	$(2a_1)^2(1b_1)^2$	1.336
SeH$_2$	$(2a_1)^2(1b_1)^2$	1.460
TeH$_2$	$(2a_1)^2(1b_1)^2$	1.658

the longer the A-H bond. The radii of the central atom AOs will be larger for (i) larger principal quantum numbers of the AOs (increasing down a column or family of the periodic table), and (ii) lower effective nuclear charge (decreasing toward the left in any row). Both of these effects can be combined in a single well-known rule: Atoms of lower electronegativity have AOs with larger radii. Table II shows typical bond distances. Lower electronegativity of A increases the size of the AOs, lengthens the A-H bonds, and lowers the energy of $2\sigma_g$. Since lower electronegativity of A also raises the energy of $1\pi_u$, the energy gap between these two MOs is reduced, causing greater mixing between the bent forms $2a_1$ and $3a_1$. More mixing gives greater stabilization to $2a_1$, which then favors smaller valence angles. The rule, then, is that as the electronegativity of A decreases through a series, the valence angle should become smaller. The following examples include both series through which the principal quantum number of the central atom A increases and series for which it remains constant while effective nuclear charge decreases.

For eight electrons in the configuration $(2a_1)^2(1b_1)^2$:

AH$_2$:	H$_2$O	H$_2$S	H$_2$Se	H$_2$Te	H$_2$F$^+$	H$_2$O	H$_2$N$^-$
HAH:	104.51°	92.06°	90.57°	90.25°	118.1°	104.51°	104°

For the seven-electron series in the ground-state configuration $(2a_1)^2(1b_1)^1$:

NH$_2$	PH$_2$	AsH$_2$	H$_2$O$^+$	H$_2$N	CH$_2^-$	H$_2$S$^+$	PH$_2$
103.3°	91.7°	90.7°	110.5°	103.3°	99°	93°	91.7°

For the seven-electron excited state configuration $(2a_1)^1(1b_1)^2$:

NH$_2$	PH$_2$
144°	123.2°

For the lowest singlet states of the six-electron series in the configuration $(2a_1)^2$:

CH$_2$	SiH$_2$	NH$_2^+$	CH$_2$	BH$_2^-$
110°	92.1°	115°	110°	102°

In the triplet states $(2a_1)^1(1b_1)^1$ the valence angles of the six-electron species are wider but still follow the same electronegativity rule:

CH$_2$	SiH$_2$
136°	123°

Finally, for the ground states of five-electron species with configuration $(2a_1)^1$:

BH$_2$	AlH$_2$
131°	119°

Recent experimental values for the energy separation or splitting between the triplet ground state and the lowest singlet state of CH$_2$ are around 8 kcal/mole (38). Laser photoelectron spectrometry of CH$_2^-$ (8) yields a CH$_2$ singlet-triplet splitting of 19.5 kcal/mole, a value that seems unreasonably large. Ab initio calculations, including extensive electron correlation effects, place this splitting at 11 kcal/mole (39). The singlet-triplet separation for NH$_2$ has not been measured, but ab initio calculations at the same level of approximation give a singlet-triplet splitting of 18 kcal/mole for CH$_2$ and 36 kcal/mole for NH$_2^+$ (30b). Although these values may be too high, one might expect their relative order to be correct, that is, the splitting in NH$_2^+$ greater than that in CH$_2$. This order is consistent with the qualitative MO model including $2a_1$-$3a_1$ mixing. The lower electronegativity of C would produce more mixing stabilization of $2a_1$ in CH$_2$ compared to NH$_2^+$, and this in turn would give greater stability to the singlet configuration of CH$_2$ and a smaller triplet-singlet separation. The singlet state should be even more stable in BH$_2^-$ than for CH$_2$, and ab initio calculations indicate that triplet BH$_2^-$ is only about 2 kcal/mole more stable than the singlet (29c). The singlet state in SiH$_2$ should also be stabilized compared to CH$_2$ because of the lower electronegativity of Si and indeed the ground state of SiH$_2$ is the singlet. The qualitative MO model is not powerful enough to predict whether the singlet or triplet state will have lower energy in any particular case, but it can explain trends in splitting energies through the series of related species.

Figure 1 can also be used to explain the shapes of AX$_2$ dihalides. According to this model, each halogen atom X contributes only a single AO to the basis set. The halogen AO might be a p orbital or some type of hybrid orbital pointing directly at the central atom A for the formation of a sigma bond. Since a ligand s AO also has the proper symmetry to form sigma bonds, the MO composition diagrams in Fig. 1 are appropriate. In counting valence electrons, include all of those from the central atom A but only one from each halogen X. For example, BeF_2 (four electrons) is linear, while BF_2 ($5e$), CF_2 ($6e$), NF_2 ($7e$), and OF_2 ($8e$) are all bent. The nine- and ten-electron species ClF_2 and ClF_2^- should be linear because the $2\sigma_g$-$3a_1$ MO system is occupied. With the exception of ClF_2, which may be bent, these conclusions are correct. In Chapter 7 the AB_2 class will be considered in detail using the full 12 AO basis set.

AH₃

Molecules of the general formula AH_3 with eight valence electrons are pyramidal or nonplanar, C_{3v}. Those with six electrons have planar centered equilateral triangular shapes, D_{3h}. Both planar and pyramidal structures are known among the seven-electron radicals. Table III lists some known or proposed AH_3 species (40–73).

Figure 4 contains qualitative AO composition diagrams for AH_3 MOs in planar and pyramidal shapes. A total of seven MOs can be formed from the seven available AOs. First, consider the MOs of planar D_{3h} geometry. The lowest energy orbital is $1a_1'$, made from the in-phase overlap of the central atom s AO with the $1s$ AOs on the hydrogens. If the central atom is rather electronegative, such as nitrogen in NH_3, the $1a_1'$ MO will be mainly the s AO of the central atom with only small perturbation contributions from

TABLE III

Shapes of known or proposed AH_3 molecules[a]

Number of valence electrons	AH_3	Shape
3	LiH_3^+	
4	HeH_3^+, BeH_3^+	Y-shaped, C_{2v}
5		
6	BH_3, AlH_3, BeH_3^-, CH_3^+, SiH_3^+, SnH_3^+	planar, D_{3h}
7	BH_3^-, CH_3, NH_3^+	
8	AlH_3^-, SiH_3, PH_3^+, GeH_3, SnH_3 NH_3, PH_3, AsH_3, SbH_3, CH_3^-, SiH_3^- H_3O^+, H_3S^+	pyramidal, C_{3v}

[a] For experimental and calculated structures, see Refs. 40–73.

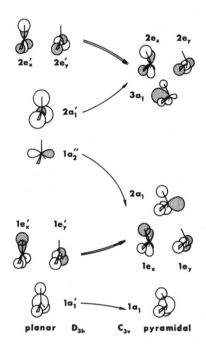

FIG. 4 Qualitative MO correlation diagram for an AH_3 molecule in planar and pyramidal shapes. (After Ref. 80.)

the hydrogens. Doubly degenerate MOs occur for molecules that have a symmetry axis of 3-fold or greater. For planar AH_3, the degenerate MOs are the pair of $1e'$ orbitals formed from the central atom p AOs that lie in the plane of the molecule and overlap in-phase with the hydrogen $1s$ AOs. The relative sizes of the hydrogen coefficients are the same as those in the e' orbitals of equilateral triangular (D_{3h}) H_3. As shown in **5** the central atom p AO in $1e_x'$ is in 100% overlap with one hydrogen $1s$ AO and in 50%

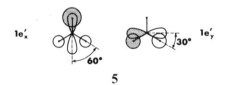

5

overlap with each of two other $1s$ AOs. In the e_y' MO the p AO overlaps each of two hydrogen $1s$ orbitals by 87% while the third hydrogen lies on the nodal plane of the MO. Overlap totals suggest similar energies for the two MOs. Recall from the H_3 example that the two $1s$ coefficients in $1e_y'$ must be larger than the three $1s$ coefficients in $1e_x'$, making the energies of the two MOs even closer. Symmetry requires that they be degenerate. Above $1e'$ is the pure p MO $1a_2''$ and finally at high energy are $2a_1'$ and $2e'$, the antibonding MOs related to the bonding orbitals $1a_1'$ and $1e'$.

Now imagine the valence angle decreasing from 120° (planar, D_{3h}) to 110° (pyramidal, C_{3v}). The overlap increases among hydrogen $1s$ AOs along $1a_1'(D_{3h})$-$1a_1(C_{3v})$, but the energy lowering is small because the hydrogen $1s$ coefficients are small. The $1s$, p overlap decreases are small along $1e'(D_{3h})$-$1e(C_{3v})$, but they raise the energy of the pyramidal structure enough to hold six-electron molecules planar since four electrons occupy this MO system. In the planar-pyramidal transformation, the three hydrogens move off the nodal surface of $1a_2''$. As they do so their $1s$ AOs can enter the resulting $2a_1$ MO and overlap one of the lobes of the central atom p AO (6).

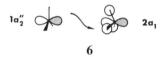

6

Since this overlap starts from zero it increases rapidly and the energy of $1a_2''$-$2a_1$ falls much more steeply than that of $1e'$-$1e$ rises below. For eight-electron molecules such as NH_3, the $1a_2''$-$2a_1$ orbital is doubly occupied and the molecules are pyramidal rather than planar. Note, however, that ab initio MO calculations show that the quantitative barrier and perhaps even the pyramidal configuration of ammonia arise only when nitrogen d orbitals are included in the basis set (69b). The qualitative features of the orbital correlation diagram (Fig. 4) remain unchanged in the detailed calculation. For the seven-electron radicals and radical ions such as CH_3, SiH_3, NH_3^+, and PH_3^+, the energy changes of fully occupied $1e'$-$1e$ and singly occupied $1a_2''$-$2a_1$ are comparable and the qualitative model cannot make definitive structure predictions. Indeed, some of these radicals are planar while others are pyramidal. Orbital mixing considerations can help us rationalize the structure trends through the seven-electron series, as well as valence angle trends and relative inversion barrier heights in the eight-electron series (74).

The $1a_2''$ and $2a_1'$ MOs of $AH_3(D_{3h})$ bear much the same relationship to each other as do the $1\pi_{ux}$ and $2\sigma_u$ MOs of $AH_2(D_{\infty h})$. D_{3h} symmetry eliminates central atom s participation in $1a_2''$ and central atom p addition to $2a_1'$. The MOs must be either symmetric or antisymmetric with respect to reflection in the plane of the molecule. This symmetry plane is lost in pyramidal geometry and both $1a_2''$ and $2a_1'(D_{3h})$ become MOs of a_1 symmetry under C_{3v}. The central atom s can add to $2a_1(C_{3v})$ and the central atom p can enter $3a_1$. This is what happens if we mix the highest energy pair of MOs of a_1 symmetry, $2a_1$ and $3a_1$, produced directly by deformation of the planar orbitals $1a_2''$ and $2a_1'$ (Fig. 5). As in the AH_2 case, the extent of

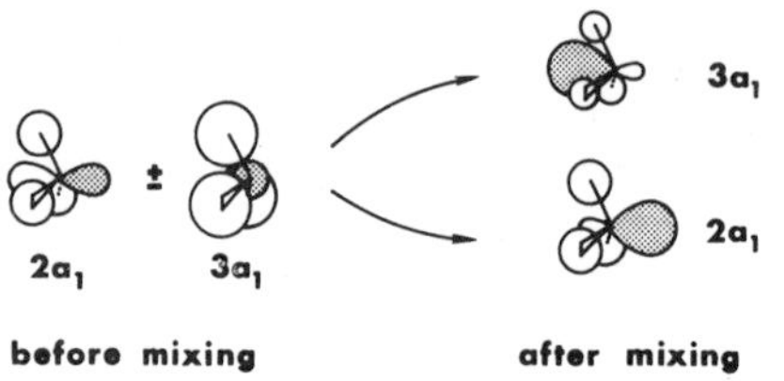

Fig. 5 Mixing of the highest energy orbital pair of a_1 symmetry in pyramidal geometry.

2a$_1$ **3a$_1$**

before mixing   **after mixing**

$2a_1$ and $3a_1$ mixing is governed by the energy gap between the unmixable $1a_2''$ and $2a_1'$ MOs of planar geometry. This spacing in turn is determined by exactly the same rules as in the AH$_2$ case: the larger the central atom and the higher the central atom AO energies, the closer are $1a_2''$ and $2a_1'$ and the more mixing stabilizes $2a_1(C_{3v})$. Therefore, we can expect AH$_3$ species with less electronegative central atoms to be more strongly pyramidal and have smaller valence angles and higher inversion barriers. For the most part, the experimental data and the results of ab initio calculations support these conclusions. Notice the trend of decreasing valence angles through the group V hydrides:

AH$_3$:	NH$_3$	PH$_3$	AsH$_3$	SbH$_3$
HAH:	106.7°	93.3°	92.1°	91.6°

For hydrides in the same row of the periodic table there is a similar decrease in valence angle with decreasing central atom electronegativity:

AH$_3$:	H$_3$O$^+$	NH$_3$	CH$_3^-$	SH$_3^+$	PH$_3$	SiH$_3^-$
HAH:	110°	106.7°	105°	96.0°	93.3°	101.3°

The bond angle in SiH$_3^-$ (from ab initio calculations) is much too large to fit the trend.

The accompanying tabulation shows that inversion barriers (in kcal/mole) of pyramidal AH$_3$ species roughly follow the rule that lower electronegativity of A produces a higher inversion barrier:

CH$_3^-$	NH$_3$	H$_3$O$^+$
5.3–8	5.8	2
	PH$_3$	H$_3$S$^+$
	30–36	32–35
	AsH$_3$	H$_3$Se$^+$
	46	30

Among the seven-electron AH$_3$ radicals, those with less electronegative central atoms are pyramidal (AlH$_3^-$, SiH$_3$, GeH$_3$, SnH$_3$, and PH$_3^+$) while those with more electronegative central atoms are planar (BH$_3^-$,

CH_3, $NH_3{}^+$). The ESR spectra of $BH_3{}^-$, CH_3, and $NH_3{}^+$ are all similar and have been interpreted as being due to planar structures. Analyses of the ultraviolet spectrum (46a) and the photoelectron spectrum (46b) of $NH_3{}^+$ also indicate a planar structure for this ion. Ab initio SCF calculations for the series $NH_3{}^+$, CH_3, and $BH_3{}^-$ show progressively less rigidly planar structures, with a very shallow minimum for $BH_3{}^-$ in pyramidal geometry (62). Estimates of valence angles for the seven-electron radicals can be calculated from the central atom isotropic hyperfine coupling measured by ESR spectroscopy. These estimates show that greater deviations from planarity follow decreasing central atom electronegativity through the series $PH_3{}^+$, SiH_3, and $AlH_3{}^-$ (48a). However, similar estimates give valence angles that increase for larger central atoms through the series SiH_3, GeH_3, and SnH_3 (51), contrary to the trend expected from the electronegativity rule.

Still other geometries are accessible to AH_3 molecules. Consider the transformation among Y-shaped (C_{2v}), equilateral triangular (D_{3h}), and T-shaped (C_{2v}) structures (Fig. 6). Figure 7 correlates the MOs of these

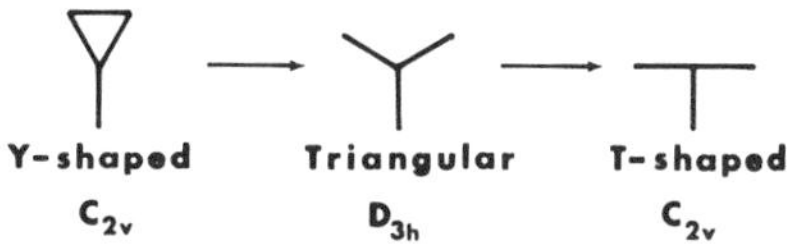

FIG. 6   T and Y shapes for AH_3.   **Y-shaped** **Triangular** **T-shaped**
C_{2v} D_{3h} C_{2v}

three structures. Since this diagram intersects the planar (D_{3h})-pyramidal (C_{3v}) diagram of Fig. 4, energy comparisons for several different geometries are immediately possible. The energy of $1a_1$ (Y)-$1a_1{}'$ (D_{3h})-$1a_1$ (T) is roughly constant for a molecule like ammonia. The energy of $1b_1$ (Y)-$1a_2''$ (D_{3h})-$1b_1$ (T) is exactly constant since the hydrogen atoms move on the nodal surface of the central atom p AO and no overlap changes are possible. Distortions to T and Y shapes break up the degeneracy of both e' (D_{3h}) MO sets, each e' pair splitting into MOs of a_1 and b_2 symmetry under C_{2v}. From $1e'$ (D_{3h}) to $1b_2$ (T) the hydrogen $1s$ AOs move into the most favorable overlap positions on the axis of the central atom p, lowering the energy of $1b_2$ (T) relative to $1e'$ (D_{3h}). At the same time the energy of $2a_1$ (T) rises as the two hydrogen $1s$ AOs approach the nodal surface of the MO. Since the overlap changes along $1e'$-$1b_2$ (T) are small (from 87 to 100%) compared to those on $1e'$-$2a_1$ (T) (50 to 0%), the distortion to T shape produces a net energy increase compared to planar triangular geometry for molecules with five, six, seven, or eight electrons. Similar arguments show that Y-shaped distortions also increase the energies of five, six, seven, or eight electron molecules.

Figure 7 indicates that three- and four-electron AH_3 molecules should be either T or Y shaped. The energies of $2a_1$ (Y) and $1b_2$ (T) are close but

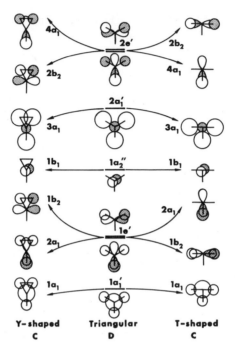

Y-shaped   Triangular   T-shaped
C   D   C

FIG. 7 Qualitative orbital correlations between planar triangular and T and Y shapes for AH₃. (After Ref. 80.)

notice that in $2a_1$ (Y) two hydrogen $1s$ AOs can overlap each other as well as a lobe of the central atom p. This better overlap makes the Y shape preferred for three- and four-electron molecules. Ab initio calculations predict HeH_3^+, LiH_3^+, and BeH_3^+ to be Y shaped. However, the structures of these ions differ in an interesting way. In LiH_3^+ and BeH_3^+, the Li and Be atoms occupy the central position, while in HeH_3^+ the He atom is a substituent on the handle or stem of the Y (7). The difference can be rationalized using

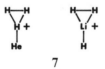

7

perturbation arguments and electronegativity considerations. The valence AO of helium is its $1s$ orbital of very deep energy. The helium $2p$ AO, the only p orbital available to make $2a_1$ (Y), is high in energy. With no p AO to hold $2a_1$ (Y) together, this MO is nonbonding in HeH_3^+. The question is whether such a nonbonding MO would be more stable with helium or a hydrogen on the stem of the Y. Since the helium AO has a lower energy

(-54.4 eV) than the hydrogen AO (-13.6 eV), it would be better to locate the He $1s$ AO away from the node and in a place where it can participate in $2a_1$ (Y) (8). The situation is just the opposite in LiH_3^+ and BeH_3^+. The

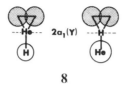

8

Li and Be atoms have low electronegativities and their valence $2s$ and $2p$ AOs have high energy relative to a hydrogen $1s$. In these cases we put the lower energy hydrogen $1s$ on the stem of the Y and the higher energy Li or Be atom at the center where the $2p$ AO can participate in the bonding.

Consider the higher energy MOs related to the antibonding $2a'$ and $2e'$ orbitals of planar triangular geometry. Notice that both T and Y shapes have a pair of $3a_1$ and $4a_1$ MOs. Mixing this highest energy pair of orbitals of a_1 symmetry will stabilize $3a_1$ and destabilize $4a_1$ in each case. Since $3a_1$ and $4a_1$ are closer together in T shape, the mixing stabilization of $3a_1$ (T) will be greater than that of $3a_1$ (Y) (Fig. 8). In fact, the $3a_1$ and $4a_1$ MOs of Y shape do very little mixing. Either combination has the sp hybrid lobe pointing directly at either one or two hydrogens, a high-energy arrangement. In the after-mixing form of $3a_1$ (T) the hybrid lobe points away from all substituents. No comparable conformation is possible for $3a_1$ (Y). A ten-electron molecule would prefer the T-shaped structure.

The correlation diagrams of Figs. 4 and 7 can be used to rationalize the shapes of AB_3 trihalides. As in the similar model of AB_2 dihalides, count all valence electrons from the central atom and only one from each halogen

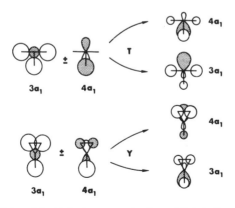

FIG. 8 Mixing of the highest energy pair of a_1 orbitals in T and Y shapes.

substituent. The six-electron trihalides (BF_3, CF_3^+) should be planar, eight-electron molecules (NF_3, $SnCl_3^-$, SF_3^+) should be pyramidal, and the ten-electron species (ClF_3, $SeCl_3^-$, XeF_3^+) should be T shaped, in agreement with experimental structures. The AB_3 series is studied in detail with a full 16 AO basis set in Chapter 7.

AH_4

The tetrahedral structure of methane can be explained without assuming tetrahedral hybrid orbitals on carbon. Figure 9 displays the lower energy valence MOs for tetrahedral (T_d), square planar (D_{4h}), and square pyramidal (C_{4v}) structures. The lowest energy or nodeless orbital $1a_1(T_d)$-$1a_{1g}(D_{4h})$-$1a_1(C_{4v})$ does not affect the shapes of molecules to be considered here. Tetrahedral symmetry permits a set of triply degenerate MOs and these are formed from the central atom p orbitals. Degenerate MOs can be represented in an infinite number of equivalent ways. The set pictured in Fig. 10 is obviously degenerate, but they are not the most convenient representations for the arguments to be presented here. Therefore, a different but equivalent set is contained in Fig. 9.

In t_x and t_y (Fig. 9) each of two $1s$ AOs is in 82% ($=100 \times \cos 35°$) of maximum possible overlap with the central atom p AO. In t_z, each of four $1s$ AOs overlap the central atom p by 57% ($=\cos 55°$) of maximum.

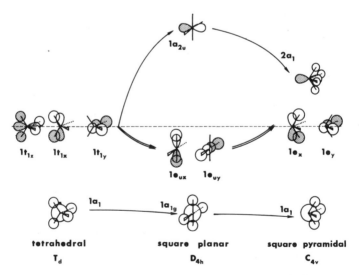

FIG. 9   Qualitative orbital correlations for AH_4 comparing tetrahedral, square planar, and square pyramidal shapes. (After Ref. 80.)

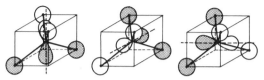

FIG. 10 An alternative and equivalent representation of the triply degenerate t MOs of AH₄ in tetrahedral geometry.

Although the overlaps in t_z ($4 \times 57 = 228$) are greater than those in t_x and t_y ($2 \times 82 = 164$), the energies are the same. Recall that, because there are more AOs in t_z, MO normalization requires coefficients of the $1s$ AOs in t_z to be smaller than those in t_x or t_y. Energies depend on the size of the AO coefficients as well as the overlaps.

The MOs for square planar geometry can be readily deduced from those for the tetrahedron by noting changes in AO composition and overlaps as the tetrahedron is flattened into the vertical plane perpendicular to the page. In the tetrahedral orbitals $1t_x$ and $1t_y$, the hydrogen AOs are already in good overlap with the lobes of the central atom p orbital. In square planar geometry these overlaps are even better, though the increases are small (82 to 100% of maximum) yielding a modest energy lowering. Along $1t_z(T_d)$-$1a_{2u}(D_{4h})$, the overlap decrease is larger (55 to 0%) producing a larger energy rise. For molecules with eight valence electrons, such as BH_4^-, CH_4, and NH_4^+, the high energy of $1a_{2u}(D_{4h})$ prohibits square planar geometry.

Next, consider the C_{4v} pyramidalization of the square planar shape. To simplify energy comparisons, assume a tetrahedral angle between hydrogens across the diagonal of the square base of the pyramid (**9**). Because of

9

identical AO composition and overlaps, the energies of e (C_{4v}) must be the same as the $1t_x$ and $1t_y$ MOs of tetrahedral geometry. In nonplanar geometry, hydrogen atoms are no longer on the nodal plane of $1a_{2u}(D_{4h})$. In $2a_1(C_{4v})$ the hydrogen $1s$ AOs are in positions where they can overlap with one lobe of the central atom p. There are the same number of $1s$, p overlaps in $2a_1(C_{4v})$ and $1t_z(T_d)$ and the overlaps have the same magnitude but the energies of these two MOs are not equal because the AO coefficients are not the same. In $2a_1(C_{4v})$ all four hydrogen $1s$ AOs are on the same side of the nodal surface and they enter the MO with the same phase. Therefore MO normalization requires that the $1s$ coefficients in $2a_1(C_{4v})$ be smaller than those in

$1t_z(T_d)$, which has more phase differences among its $1s$ AOs. With smaller coefficients the orbital energy of $2a_1(C_{4v})$ will be higher than that of $1t_z(T_d)$.

A comparison of energies for eight-electron molecules in tetrahedral, square planar, and square pyramidal shapes shows that tetrahedral geometry is the most stable. However, this comparison gives the surprising result that the square pyramid should have lower energy than the square planar shape, a conclusion supported by the results of ab initio calculations (77). Qualitative considerations do not say how high the square pyramid is above the tetrahedron, but they do suggest that those who attempt to prepare alternative forms of tetracoordinate carbon might find the square pyramid a more easily realizable goal.

The relatively low energy of the e_u (D_{4h}) MOs in Fig. 9 indicates that AH$_4$ molecules with six-valence electrons, such as the hypothetical BeH$_4$, should be square planar rather than tetrahedral. The results of recent ab initio MO calculations for the six-electron series BeH$_4$, BH$_4^+$, and CH$_4^{2+}$ show that these species do indeed have lower energy in the square planar geometry (78). However, those studies found BeH$_4$ to be unstable with respect to BeH$_2$ and H$_2$ and that CH$_4^{2+}$ dissociates into the lower energy products CH$_3^+$ and H$^+$. Stable structures of C_{2v} symmetry exist for BH$_4^+$ consisting of a slightly bent BH$_2^+$ unit loosely associated with H$_2$. These H$_2$B$^+ \cdots$ H$_2$ structures have nearly equal energy and are only a couple of kcal/mole below the energy of separated BH$_2^+$ and H$_2$. This seemingly bewildering variety of results for rather similar systems can be understood with the aid of qualitative MO ideas plus simple electrostatic arguments.

The six-electron AH$_4$ molecules should be square planar and four-electron AH$_2$ molecules should be linear. Figure 11 is the correlation diagram for dissociation of square planar AH$_4$ into linear AH$_2$ and H$_2$. The representation of the degenerate e_u (D_{4h}) MOs in Fig. 11 is different from but equivalent to that in Fig. 9. The relationship between the two sets of MOs is simple. The normalized sum of two p AOs on the same atom is another p AO that lies along the diagonal between the two summed AOs (**10**).

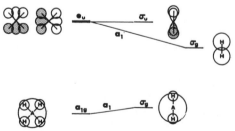

FIG. 11   Orbital correlations for the dissociation AH$_4$(D_{4h}) $\xrightarrow{C_{2v}}$ AH$_2$($D_{\infty h}$) + H$_2$. (Reprinted from Ref. 81 by permission of Pergamon Press, Inc.)

10

Assuming that the dissociation follows C_{2v} symmetry, the process is allowed. Electrons flow smoothly from reactant energy levels to fill the appropriate energy levels of the correct products. The σ_u orbital of linear AH$_2$ has exactly the same energy as the e_u pair of square planar AH$_4$. Compare σ_u(AH$_2$) in Fig. 11 and $1e_{ux}$(AH$_4$) in Fig. 9. The nodeless σ_g MO of AH$_2$ would be slightly higher but otherwise comparable in energy to the nodeless a_{1g} MO of AH$_4$. Whether the dissociation AH$_4$ → AH$_2$ + H$_2$ is exothermic or endothermic depends on the relative energies of σ_g(H$_2$) and σ_u(AH$_2$). The energy of σ_g(H$_2$) will be the same whether the reactant is BeH$_4$, BH$_4{}^+$, or CH$_4^{2+}$. The energy of σ_u(AH$_2$) depends on the energy of the $2p$ AOs of the central atom A. For Be, with low electronegativity and high-energy $2p$ AOs, the energy of σ_u(AH$_2$) is higher than that of σ_g(H$_2$) and the dissociation of BeH$_4$ into BeH$_2$ and H$_2$ would be exothermic as the ab initio calculations indicate and as implied by the order of product energy levels in Fig. 11. Boron and carbon have progressively larger electronegativities and lower energy $2p$ AOs and these move the σ_u(AH$_2$)-e_u(AH$_4$) level to lower energies. For BH$_4{}^+$, the energies of σ_g(H$_2$) and σ_u(AH$_2$) must be about the same to account for the loosely bound H$_2$B$^+$ $\cdots$ H$_2$ complex. For CH$_4^{2+}$, σ_u(AH$_2$) must be lower than σ_g(H$_2$), making the reaction CH$_4^{2+}$ → CH$_2^{2+}$ + H$_2$ endothermic, blocking the dissociation. Instead CH$_4^{2+}$ breaks into CH$_3{}^+$ and H$^+$. But why does CH$_4^{2+}$ give up H$^+$ while BeH$_4$ and BH$_4{}^+$ do not? Figure 12 is the correlation diagram for the dissociation of square planar AH$_4$ into planar triangular AH$_3$ and H. Again, assume a process of C_{2v} symmetry. The hydrogen atom nonbonding $1s$ AO should be far above the energies of the bonding MOs of AH$_4$ and AH$_3$. All six electrons of AH$_4$ flow into the AH$_3$

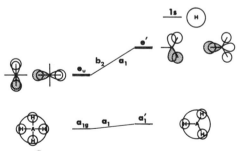

FIG. 12   AH$_4$(D_{4h}) $\xrightarrow{\ C_{2v}\ }$ AH$_3$(D_{3h}) + H. (Reprinted from Ref. 81 by permission of Pergamon Press, Inc.)

product orbitals, leaving the hydrogen $1s$ AO empty and producing H$^+$. Hydrogen $1s$ and A atom $2p$ AO overlap differences between $e_u(\text{AH}_4)$, and $e'(\text{AH}_3)$ clearly show that the e' levels of product are above the energy of the related e_u levels of reactant, making the dissociation endothermic for any A and blocking the process. This conclusion is correct for BeH$_4$ and BH$_4^+$ but the dissociation of CH$_4^{2+}$ occurs anyway. Why?

The qualitative MO model and the extended Hückel method on which the qualitative model is based fail to account properly for changes in electrostatic potential. Internal electrostatic repulsions clearly increase through the series BeH$_4$, BH$_4^+$, and CH$_4^{2+}$. The dissociation of BeH$_4$ into BeH$_3^-$ and H$^+$ would separate opposite charges. Both MO and electrostatic energetics oppose this process. The dissociation of BH$_4^+$ into BH$_3$ and H$^+$ would produce no new ions but it would lower internal electrostatic repulsions in BH$_4^+$. Electrostatic repulsions in CH$_4^{2+}$ could be substantially reduced by dissociation into two ions: CH$_3^+$ and H$^+$. This electrostatic driving force for dissociation of H$^+$, reversed in direction in BeH$_4$ and small or zero in BH$_4^+$, is large in CH$_4^{2+}$, where it apparently outweighs the qualitative MO energetics. Similar considerations are not appropriate for the processes summarized in Fig. 11 because the product projected in each case is neutral H$_2$.

Other shapes are possible for AH$_4$ molecules. One is the elongation or stretching of the regular tetrahedron to D_{2d} geometry. The correlation diagram for this process shows that four-electron species, perhaps H$_5^+$ or LiH$_4^+$, have minimum energy in the D_{2d} shape (**11**). This is due to the

11

increased in-phase overlaps and reduced out-of-phase overlaps in the upper of the two occupied MOs. The lower occupied orbital is the nodeless a_1 MO composed of in-phase overlapping s AOs. In LiH$_4^+$, the central Li atom can contribute a $2p$ AO to $1b_2(D_{2d})$ to stabilize a symmetric structure (57). No $2p$ orbital is readily available from hydrogen, however, and this may account for the fact that ab initio calculations predict that the structure of H$_5^+$ is an elongated but distorted tetrahedron (C_{2v} rather than D_{2d}), essentially an H$_3^+$ ion loosely bound to an H$_2$ molecule (79) (**12**). The less

12

symmetric structure allows the central hydrogen to move off of the nodal surface of the higher occupied MO.

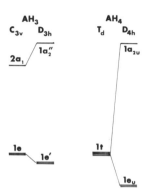

FIG. 13   Comparison of higher energy valence levels of AH_3 in pyramidal and planar shapes and for AH_4 in tetrahedral and square planar shapes. The relatively small gap between $2a_1$ and $1a_2''$ permits inversion of eight-electron AH_3 molecules. Inversion of eight-electron AH_4 molecules is much more difficult because of the large gap between t_1 and a_u. (From Ref. 80.)

INVERSION

Qualitative MO pictures can show why ammonia inverts but methane does not. The most straightforward comparison for this purpose would be between NH_3 and $NH_4{}^+$. Figure 13 contains energy levels in pyramidal (C_{3v}) and planar triangular (D_{3h}) shapes for AH_3 compared with those for AH_4 in tetrahedral (T_d) and square planar (D_{4h}) shapes. Figures 4 and 9 form the basis for Fig. 13. The a_{2u} level of planar AH_4 is prohibitively high in energy relative to the t orbitals of tetrahedral geometry, while the $2a_1$ orbital of pyramidal AH_3 is much closer to the $1a_2''$ of planar shape, making the inversion of AH_3 much cheaper energetically.

REFERENCES

Experimental AH_2 *Structures or Other Data*

(1) BH_2: G. Herzberg and J. W. C. Johns, *Proc. R. Soc. London* **A298**, 142 (1967).

(2) AlH_2: G. Herzberg and J. W. C. Johns, quoted in G. Herzberg, "Molecular Spectra and Structure," Vol. III, Electronic Spectra and Electronic Structure of Polyatomic Molecules, p. 583. Van Nostrand-Reinhold, Princeton, New Jersey, 1966.

(3) CH_2: G. Herzberg and J. W. C. Johns, *J. Chem. Phys.* **54**, 2276 (1971); *Proc. R. Soc. London* **A295** (1966); R. A. Bernheim, H. W. Bernard, P. S. Wang, L. S. Wood, and P. S. Skell, *J. Chem. Phys.*, **54**, 3223 (1971).

(4) SiH_2: I. Dubois, *Can. J. Phys.* **46**, 2485 (1968); I. Dubois, G. Duxbury, and R. N. Dixon, *J. Chem. Soc. Faraday II* **71**, 799 (1975); O. F. Zeck, Y. Y. Su, G. P. Gennaro, and Y.-N. Tang, *J. Am. Chem. Soc.* **96**, 5967 (1974).

(5) GeH_2: G. R. Smith and W. A. Guillory, *J. Chem. Phys.* **56**, 1423 (1972).

(6) NH_2: K. Dressler and D. A. Ramsay, *Phil. Trans. Roy. Soc.* (*London*) **A251**, 553 (1959); R. N. Dixon, *Mol. Phys.* **9**, 357 (1965).

(7) PH$_2$: J. M. Berthou, B. Pascat, H. Guenebaut, and D. A. Ramsay, *Can. J. Phys.* **50**, 2265 (1972).

(8) CH$_2^-$: P. F. Zittel, G. B. Ellison, S. V. O'Neil, E. Herbst, W. C. Lineberger, and W. P. Reinhardt, *J. Am. Chem. Soc.* **98**, 3731 (1976).

(9) SiH$_2^-$: A. Kasdan, E. Herbst, and W. C. Lineberger, *J. Chem. Phys.* **62**, 541 (1975).

(10) AsH$_2$: R. N. Dixon, G. Duxbury, and H. M. Lamberton, *Proc. R. Soc. London* **A305**, 271 (1968).

(11) H$_2$O$^+$: H. Lew and I. Heiber, *J. Chem. Phys.* **58**, 1246 (1973).

(12) H$_2$S$^+$: G. Duxbury, M. Horani, and J. Rostas, *Proc. R. Soc. London* **A331**, 109 (1972).

(13) H$_2$O: R. L. Cook, F. C. DeLucia, and P. Helminger, *J. Mol. Spectrosc.* **53**, 62 (1974).

(14) H$_2$S: T. H. Edwards, N. K. Moncur, and L. E. Snyder, *J. Chem. Phys.* **46**, 2139 (1967); A. R. Gallo and K. K. Innes, *J. Mol. Spectrosc.* **54**, 472 (1975); R. L. Cook, F. C. DeLucia, and P. Helminger, *J. Mol. Struct.* **28**, 237 (1975).

(15) H$_2$Se: R. A. Hill and T. H. Edwards, *J. Chem. Phys.* **42**, 1391 (1965).

(16) H$_2$Te: N. K. Moncur, P. D. Willson, and T. H. Edwards, *J. Mol. Spectrosc.* **52**, 380 (1974).

(17) NH$_2^-$: S. F. Mason, *J. Phys. Chem.* **61**, 384 (1957).

(18) PH$_2^-$: P. F. Zittel and W. C. Lineberger, *J. Chem. Phys.* **65**, 1236 (1976).

(19) FH$_2^+$: M. Couzi, J -C. Cornut, and P. V. Huong, *J. Chem. Phys.* **56**, 426 (1972).

Calculated AH$_2$ *Structures*

(20) LiH$_2^+$: R. C. Raffenetti and K. Ruedenberg, *J. Chem. Phys.* **59**, 5978 (1973); W. Kutzelnigg, V. Staemmler, and C. Hoheisel, *Chem. Phys.* **1**, 27 (1973).

(21) LiH$_2$: A. L. Companion, *J. Chem. Phys.* **48**, 1186 (1968).

(22) BeH$_2^+$: R. D. Poshusta, D. W. Klint, and A. Liberles, *J. Chem. Phys.* **55**, 252 (1971).

(23) BeH$_2$: S. D. Peyerimhoff, R. J. Buenker, and L. C. Allen, *J. Chem. Phys.* **45**, 734 (1966); J. J. Kaufman, L. M. Sachs, and M. Geller, *J. Chem. Phys.* **49**, 4369 (1968).

(24) BH$_2^+$: S. D. Peyerimhoff, R. J. Buenker, and L. C. Allen, *J. Chem. Phys.* **45**, 734 (1966); M. Jungen, *Chem. Phys. Lett.* **5**, 241 (1970).

(25) BH$_2^-$: C. F. Bender and H. F. Schaefer III, *J. Mol. Spectrosc.* **37**, 423 (1971).

(26) CH$_2^+$: C. F. Bender and H. F. Schaefer III, *J. Mol. Spectrosc.* **37**, 423 (1971).

(27) CH$_2$: J. F. Harrison and L. C. Allen, *J. Am. Chem. Soc.* **91**, 807 (1969); J. E. DelBene, *Chem. Phys. Lett.* **9**, 68 (1971); S. V. O'Neil, H. F. Schaefer, III, and C. F. Bender, *J. Chem. Phys.* **55**, 162 (1971); C. F. Bender, H. F. Schaefer, III, D. R. Franceschetti, and L. C. Allen, *J. Am. Chem. Soc.* **94**, 6888 (1972); S. Y. Chu, A. K. Q. Siu, and E. F. Hayes, *J. Am. Chem. Soc.* **94**, 2969 (1972).

(28) CH$_2^-$: R. B. Davidson and M. L. Hudak, *J. Am. Chem. Soc.* **99**, 3918 (1977).

(29) BH$_2^-$: (a) S. D. Peyerimhoff, R. J. Buenker, and L. C. Allen, *J. Chem. Phys.* **45**, 734 (1966); (b) L. M. Sachs, M. Geller, and J. J. Kaufman, *J. Chem. Phys.* **52**, 974 (1970); (c) L. Z. Stenkamp and E. R. Davidson, *Theoret. Chim. Acta* **30**, 283 (1973).

(30) NH$_2^+$: (a) S. T. Lee and K. Morokuma, *J. Am. Chem. Soc.* **93**, 6863 (1971); (b) S. Y. Chu, A. K. Q. Siu, and E. F. Hayes, *J. Am. Chem. Soc.* **94**, 2969 (1972).

(31) SiH$_2$: B. Wirsam, *Chem. Phys. Lett.* **14**, 214 (1972).

(32) H$_2$O$^+$, H$_2$S$^+$: H. Sakai, S. Yamaba, T. Yamaba, and K. Fukui, *Chem. Phys. Lett.* **25**, 541 (1974).

(33) H$_2$O: T. H. Dunning, Jr., R. M. Pitzer, and S. Aung, *J. Chem. Phys.* **57**, 5044 (1972); D. R. McLaughlin, C. F. Bender, and H. F. Schaefer, III, *Theoret. Chim. Acta* **25**, 352 (1973).

(34) H$_2$S: B. Cadioli, U. Pincelli, G. L. Bendazzoli, and P. Palmieri, *Theoret. Chim. Acta* **19**, 66 (1970).

(35) H$_2$F$^+$: H. von Hirschhausen, D. F. Ilten, and E. Zeeck, *Chem. Phys. Lett.* **40**, 80 (1976).

General References

(36) G. A. Gallup, *J. Chem. Phys.* **26**, 716 (1957); R. Hoffmann, G. D. Zeiss, and G. W. Van Dine, *J. Am. Chem. Soc.* **90**, 1485 (1968).

(37) C. R. Brundle and D. W. Turner, *Proc. R. Soc. London* **A307**, 27 (1968).

(38) H. M. Frey, *Chem. Commun.* 1024 (1972); J. W. Simons and R. Curry, *Chem. Phys. Lett.* **38**, 171 (1976).

(39) L. B. Harding and W. A. Goddard, III, *J. Chem. Phys.* **67**, 1777 (1977); R. R. Lucchese and H. F. Schaefer, III, *J. Am. Chem. Soc.* **99**, 6765 (1977).

Experimental AH$_3$ Structures and Other Data

(40) BeH$_3{}^+$: S. S. Brenner and S. R. Goodman, *Nature (London) Phys. Sci.* **235**, 35 (1972).

(41) BH$_3$: P. S. Ganguli and H. A. McGee, *J. Chem. Phys.* **50**, 4658 (1969); G. W. Mappes and T. P. Fehlner, *J. Am. Chem. Soc.* **92**, 1562 (1970); A. Kaldor and R. F. Porter, *J. Am. Chem. Soc.* **93**, 2140 (1971).

(42) AlH$_3$: P. Breisacher and B. Siegel, *J. Am. Chem. Soc.* **86**, 5053 (1964).

(43) SnH$_3{}^+$: J. R. Webster and W. F. Jolly, *Inorg. Chem.* **10**, 877 (1971).

(44) BH$_3{}^-$: M. C. R. Symons and H. W. Wardale, *Chem. Commun.* 753 (1967); R. C. Catton, M. C. R. Symons, and H. W. Wardale, *J. Chem. Soc. A* 2622 (1969); E. D. Sprague and F. Williams, *Mol. Phys.* **20**, 375 (1970).

(45) CH$_3$: G. Herzberg, *Proc. R. Soc. London* **A262**, 291 (1961); R. W. Fessenden, *J. Phys. Chem.* **71**, 74 (1967); L. Andrews and G. C. Pimentel, *J. Chem. Phys.* **47**, 3637 (1965); D. E. Milligan and M. E. Jacox, *J. Chem. Phys.* **47**, 5146 (1967); L. Y. Tan, A. M. Winer, and G. C. Pimentel, *J. Chem. Phys.* **57**, 4028 (1972).

(46) NH$_3{}^+$: (a) A. D. Walsh and P. A. Warsop, *Trans. Faraday Soc.* **57**, 345 (1961); (b) G. R. Branton, D. C. Frost, T. Makita, C. A. McDowell, and I. A. Stenhouse, *Phil. Trans. Roy. Soc. (London)* **A263**, 77 (1970).

(47) AlH$_3{}^-$: R. C. Catton and M. C. R. Symons, *J. Chem. Soc. A* 2001 (1969).

(48) PH$_3{}^+$: (a) A. Begum, A. R. Lyons, and M. C. R. Symons, *J. Chem. Soc. A* 2290 (1971); (b) J. P. Maier and D. W. Turner, *J. Chem. Soc. Faraday II* **68**, 711 (1972).

(49) SiH$_3$: G. S. Jackel and W. Gordy, *Phys. Rev.* **176**, 443 (1968); D. E. Milligan and M. E. Jacox, *J. Chem. Phys.* **52**, 2594 (1970).

(50) GeH$_3$: G. S. Jackel and W. Gordy, *Phys. Rev.* **176**, 443 (1968); G. R. Smith and W. A. Guillory, *J. Chem. Phys.* **56**, 1423 (1972).

(51) SnH$_3$: G. S. Jackel and W. Gordy, *Phys. Rev.* **176**, 443 (1968).

(52) NH$_3$: W. S. Benedict and E. K. Plyler, *Can. J. Phys.* **35**, 1235 (1957); K. Kuchitsu, J. P. Guillory, and L. S. Bartell, *J. Chem. Phys.* **49**, 2488 (1968).

(53) PH$_3$: A. G. Maki, R. L. Sams, and W. B. Olson, *J. Chem. Phys.* **58**, 4502 (1973); D. A. Helms and W. Gordy, *J. Mol. Spectrosc.* **66**, 206 (1977).

(54) AsH$_3$: W. B. Olson, A. G. Maki, and R. L. Sams, *J. Mol. Spectrosc.* **55**, 252 (1975); P. Helminger, E. L. Beeson, Jr., and W. Gordy, *Phys. Rev.* **A3**, 122 (1971).

(55) SbH$_3$: P. Helminger, E. L. Beeson, Jr., and W. Gordy, *Phys. Rev.* **A3**, 122 (1971).

(56) H$_3$O$^+$: J.-O. Lundgren and J. M. Williams, *J. Chem. Phys.* **58**, 788 (1973).

Calculated AH$_3$ Structures

(57) LiH$_3{}^+$: R. D. Poshusta, J. A. Haugen, and D. F. Zetik, *J. Chem. Phys.* **51**, 3343 (1969).

(58) HeH$_3{}^+$: R. D. Poshusta and V. P. Agrawal, *J. Chem. Phys.* **59**, 2477 (1973).

(59) BeH$_3{}^+$: M. Jungen and R. Ahlrichs, *Mol. Phys.* **28**, 367 (1974).

(60) BH$_3$, CH$_3{}^+$, BeH$_3{}^-$: S. D. Peyerimhoff, R. J. Buenker, and L. C. Allen, *J. Chem. Phys.* **45**, 734 (1966).

(61) BeH$_3$: M. E. Schwartz and L. C. Allen, *J. Am. Chem. Soc.* **92**, 1466 (1970).

(62) BH$_3$$^-$, CH$_3$, NH$_3$$^+$: T. A. Claxton and N. A. Smith, *J. Chem. Phys.* **52**, 4317 (1970).

(63) CH$_3$: K. Morokuma, L. Pedersen, and M. Karplus, *J. Chem. Phys.* **48**, 4801 (1968); R. McDiarmid, *Theoret. Chim. Acta* **20**, 282 (1971); Y. Ishikawa and R. C. Bining, Jr., *Chem. Phys. Lett.* **40**, 342 (1976).

(64) CH$_3$$^+$, CH$_3$, CH$_3$$^-$: (a) P. Millie and G. Berthier, *Int. J. Quantum Chem. Symp.* **2**, 67 (1968); (b) R. E. Kari and I. G. Csizmadia, *J. Chem. Phys.* **50**, 1443 (1969); (c) F. Dreissler, R. Ahlrichs, V. Staemmler, and W. Kutzelnigg, *Theoret. Chim. Acta* **30**, 315 (1973); (d) E. D. Jemmis, V. Buss, P. v. R. Schleyer, and L. C. Allen, *J. Am. Chem. Soc.* **98**, 6483 (1976); G. T. Surratt and W. A. Goddard, III, *Chem. Phys.* **23**, 39 (1977).

(65) CH$_3$$^-$: A. J. Duke, *Chem. Phys. Lett.* **21**, 275 (1973).

(66) CH$_3$, NH$_3$$^+$, CH$_3$$^-$, NH$_3$: R. E. Kari and I. G. Csizmadia, *Int. J. Quantum Chem.* **6**, 401 (1972); *J. Chem. Phys.* **56**, 4337 (1972).

(67) SiH$_3$$^+$, SiH$_3$, SiH$_3$$^-$: B. Wirsam, *Chem. Phys. Lett.* **18**, 578 (1973).

(68) SiH$_3$, PH$_3$$^+$, PH$_3$: L. J. Aarons, M. F. Guest, M. B. Hall, and I. H. Hillier, *J. Chem. Soc. Dalton II* **69**, 643 (1973); L. J. Aarons, I. H. Hillier, and M. F. Guest, *J. Chem. Soc. Faraday II* **70**, 167 (1974).

(69) NH$_3$: (a) R. M. Stevens, *J. Chem. Phys.* **55**, 1725 (1971); **61**, 2086 (1974); (b) A. Rauk, L. C. Allen, and E. Clementi, *J. Chem. Phys.* **52**, 4133 (1970).

(70) PH$_3$: J. M. Lehn and B. Munsch, *Mol. Phys.* **23**, 91 (1972); R. Ahlrichs, F. Keil, H. Lischka, W. Kutzelnigg, and V. Staemmler, *J. Chem. Phys.* **63**, 455 (1975).

(71) H$_3$O$^+$: (a) J. Almlöf and U. Wahlgren, *Theoret. Chim. Acta* **28**, 161 (1973); (b) P. A. Kollman and C. F. Bender, *Chem. Phys. Lett.* **21**, 271 (1973); (c) M. Allavena and E. LeClech, *J. Mol. Struct.* **22**, 265 (1974).

(72) H$_3$S$^+$: T. Yamabe, T. Aoyagi, S. Nagata, H. Sakai, and K. Fukui, *Chem. Phys. Lett.* **28**, 182 (1974); D. A. Dixon and D. S. Marynick, *J. Am. Chem. Soc.* **99**, 6101 (1977).

(73) PH$_3$, SH$_3$$^+$, AsH$_3$, SeH$_3$$^+$: D. A. Dixon and D. S. Marynick, *J. Am. Chem. Soc.* **99**, 6101 (1977); D. S. Marynick and D. A. Dixon, *Disc. Faraday Soc.* **62**, 47 (1977).

General References

(74) C. C. Levin, *J. Am. Chem. Soc.* **97**, 5649 (1975); W. Cherry and N. Epiotis, *J. Am. Chem. Soc.* **98**, 1135 (1976).

(75) J. D. Swalen and J. A. Ibers, *J. Chem. Phys.* **36**, 1914 (1962).

(76) R. E. Weston, Jr., *J. Am. Chem. Soc.* **76**, 2645 (1954).

(77) I. Shavitt, quoted by J. B. Collins, J. D. Dill, E. D. Jemmis, Y. Apeloig, P. v. R. Schleyer, R. Seeger, and J. A. Pople, *J. Am. Chem. Soc.* **98**, 5419 (1976); V. I. Minkin, R. M. Minyaev, and I. I. Zacharov, *Chem. Commun.* 213 (1977).

(78) J. B. Collins, P. v. R. Schleyer, J. S. Binkley, J. A. Pople, and L. Radom, *J. Am. Chem. Soc.* **98**, 3436 (1976).

(79) J.-T. J. Huang, M. E. Schwartz, and G. V. Pfeiffer, *J. Chem. Phys.* **56**, 755 (1972); W. I. Salmon and R. D. Poshusta, *J. Chem. Phys.* **59**, 4867 (1973); R. Ahlrichs, *Theoret. Chim. Acta* **39**, 149 (1975).

(80) B. M. Gimarc, *J. Am. Chem. Soc.* **93**, 593 (1971).

(81) B. M. Gimarc, *Tetrahedron Lett.*, 1862 (1977).

4 Properties of AB_4, AB_5, and AB_6 nontransition element complexes

This chapter discusses the shapes and other properties of selected complexes with general formulas AB_4, AB_5, and AB_6 in which the central atom A is of an element from the representative or main groups III through 0 of the periodic table and the ligands B are single atoms. In most of these complexes the ligands are halogens although some oxides and sulfides are known. The AO basis set consists of the valence s and p AOs of the central atom plus a single AO of sigma-bond symmetry from each ligand B. The ligand AO might be a p orbital or some kind of hybrid orbital pointing directly towards the central atom. For representational convenience the ligand AOs have been drawn as s AOs in all of the accompanying figures. The valence electrons are counted by taking all of the electrons in the s and p AOs of the neutral central atom A plus one electron from each halogen ligand (none from each oxygen or sulfur ligand) plus one electron for each negative charge on the complex as a whole. (If the complex bears a positive charge, this charge must be subtracted from the valence electron total.) A similar model for AB_2 and AB_3 halides was mentioned in Chapter 3.

STRUCTURES OF AB_4

AB_4 molecules with eight valence electrons have tetrahedral (T_d) geometry. Those with ten electrons have a structure usually referred to as that of a trigonal bipyramid with one of the three equatorial valence positions vacant or, rather, occupied by a lone pair of electrons. This description is based on the positions of directed electron pairs in the five-coordinate case of the valence shell electron pair repulsion (VSEPR) model. Since we are presenting an alternative model of chemical valence we introduce a different description of this C_{2v} structure, that of a square folded along one of its diagonals (*1*). We refer to the two coordinate positions on the fold axis as the axial positions

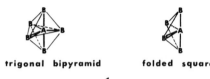

trigonal bipyramid   **folded square**

1

and the other two positions as the equatorial positions. The 12-electron AB_4 complexes are square planar (D_{4h}). Tables I, II, and III list, respectively,

TABLE I

AB_4 halides, eight electrons[a]

III	IV	V	VI	VII	0
BF_4^-	CF_4	NF_4^+			
BCl_4^-	CCl_4				
BBr_4^-	CBr_4				
BI_4^-	CI_4				
AlF_4^-	SiF_4	PF_4^+			
$AlCl_4^-$	$SiCl_4$	PCl_4^+			
$AlBr_4^-$	$SiBr_4$	PBr_4^+			
AlI_4^-	SiI_4				
GaF_4^-	GeF_4				
$GaCl_4^-$	$GeCl_4$	$AsCl_4^+$			
$GaBr_4^-$	$GeBr_4$				
GaI_4^-	GeI_4				
InF_4^-	SnF_4				
$InCl_4^-$	$SnCl_4$	$SbCl_4^+$			
$InBr_4^-$	$SnBr_4$				
InI_4^-	SnI_4				
	PbF_4				
$TlCl_4^-$	$PbCl_4$				
$TlBr_4^-$	$PbBr_4$				
TlI_4^-	PbI_4				

[a] Reprinted with permission from Gimarc and Khan, *J. Am. Chem. Soc.* **100**, 2340 (1978). Copyright by the American Chemical Society.

Table II

AB₄ halides, ten electrons[a]

III	IV	V	VI	VII	0
		PF_4^-	SF_4	ClF_4^+	
			SCl_4		
		PBr_4^-			
		AsF_4^-	SeF_4	BrF_4^+	
		$AsCl_4^-$	$SeCl_4$		
		$AsBr_4^-$	$SeBr_4$		
		SbF_4^-	TeF_4	IF_4^+	
	$SnCl_4^{2-}$	$SbCl_4^-$	$TeCl_4$		
		$SbBr_4^-$	$TeBr_4$		
		SbI_4^-	TeI_4		
	$PbCl_4^{2-}$	$BiCl_4^-$	$PoCl_4$		
	$PbBr_4^{2-}$	$BiBr_4^-$	$PoBr_4$		
TlI_4^{3-}	PbI_4^{2-}	BiI_4^-	PoI_4		

[a] Reprinted with permission from Gimarc and Khan, *J. Am. Chem. Soc.* **100**, 2340 (1978). Copyright by the American Chemical Society.

the known 8-, 10-, and 12-electron tetrahalide complexes of the representative elements in groups III through 0 of the periodic table (*1–63*). Even the vacant spaces in Tables I, II, and III contain information. To the extent that they correspond to complexes that are unknown because of their innate instability rather than high reactivity or synthetic neglect, the vacancies provide rough data concerning the relative stabilities of complexes. Later in this chapter we relate electronic structure properties to the fact that various classes of known complexes fill out different parts of such tables. All of the known tetroxides and tetrasulfides have eight valence electrons and tetrahedral geometry. Representative examples are PO_4^{3-}, SO_4^{2-}, ClO_4^-, XeO_4, SiS_4^{4-}, SnS_4^{4-}, PS_4^{3-}, AsS_4^{3-}, and SbS_4^{3-} (*64–68*).

Some nine-electron radicals have been produced in radiation damage experiments and observed by ESR spectroscopy. Examples are PO_4^{4-}, AsO_4^{4-}, SeO_4^{3-}, ClO_4^{2-}, CBr_4^-, PF_4, PCl_4, and SF_4^+ (*69–77*). These are all known or assumed to have the diagonally folded square (C_{2v}) structure of

TABLE III

AB_4 halides, twelve electrons[a]

III	IV	V	VI	VII	0
				ClF_4^-	
				BrF_4^-	
				IF_4^-	XeF_4
				ICl_4^-	
			$PoCl_4^{2-}$		

[a] Reprinted with permission from Gimarc and Khan, *J. Am. Chem. Soc.* **100**, 2340 (1978). Copyright by the American Chemical Society.

the ten-electron complexes. $POCl_3^-$ (78) and $SO_2Cl_2^-$ (79) are examples of less symmetric, mixed-ligand nine-electron radicals. These are belived to have the diagonally folded square (C_{2v}) shape with the more electronegative ligands occupying the axial positions.

The mixed-ligand ten-electron complexes also have the more electronegative ligands in the axial positions of the diagonally folded square structure (80). Examples of this class are $ClO_2F_2^-$, $IO_2F_2^-$, XeO_2F_2, XeO_3F^-, $XeOF_3^+$, and ClF_3O (81–89). There is some uncertainty about the details of the axial and equatorial substitution in $TeBr_2Cl_2$ (90,91).

The ten-electron radical ClF_4 has been observed and ESR data suggest that it is square planar D_{4h} (92).

MOLECULAR ORBITALS FOR AB_4

Starting with our eight AO basis set, the MOs of tetrahedral symmetry are two nondegenerate a_1 orbitals and two triply degenerate t_1 sets. The a_1

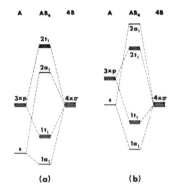

FIG. 1 Correlation of MOs for tetrahedral AB_4 with AOs for separated atoms A + 4B. (a) Electronegative central atom A. (b) Electropositive A. [Reprinted with permission from Gimarc and Khan, *J. Am. Chem. Soc.* **100**, 2340 (1978). Copyright by the American Chemical Society.]

MOs are formed by in-phase (bonding) or out-of-phase (antibonding) combinations of the ligand AOs and the central atom *s* orbital. Similarly, the t_1 sets are formed by in-phase or out-of-phase combinations of ligand AOs and individual central atom *p* orbitals. Since the central atom *s* has a lower energy than the central atom *ps*, the bonding $1a_1$ MO will lie below the bonding $1t_1$ set in energy. The consequences for molecular shapes when electrons occupy only the bonding $1a_1$ and $1t_1$ MOs have already been discussed in Chapter 3. The energy order of the antibonding orbitals $2a_1$ and $2t_1$ depends on further assumptions. Figure 1a shows the case in which the ligand AOs have energy comparable to that of the central atom *p* AOs, with a large energy gap between central atom *s* and *p* AOs. This arrangement would be suitable for a complex that has a rather electronegative central atom, ClF_4, for example. The large energy gap between central atom *s* and *p* AOs and the small perturbation interaction between ligand AOs and the central atom *s* combine to make $2a_1$ fall below $2t_1$. In Fig. 1b the ligand AO energy is near that of the central atom *s* orbital and there is a small energy gap between central atom *s* and *p* orbitals, an alignment that should be more appropriate for a complex such as AlF_4^-, in which the central atom is considerably less electronegative than the ligands. Here the small energy gap between the central atom *s* and *p* and the strong perturbation interaction between the central atom *s* and the ligand AOs combine to push $2a_1$ above $2t_1$.

Figure 2 correlates orbital energy levels for AB_4 complexes through successive angular rearrangements from regular tetrahedral (T_d) through diagonally folded square (C_{2v}) to square planar (D_{4h}) geometries, assuming that all A-B bond lengths remain equal and constant. The order of energy levels chosen for tetrahedral geometry is that for an electronegative central atom, Fig. 1a.

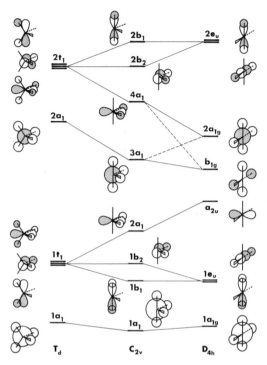

FIG. 2   MO correlations of AB_4 complexes in tetrahedral (T_d), diagonally folded square (C_{2v}), and square planar shapes, assuming the ordering of tetrahedral levels from Fig. 1a. [Reprinted with permission from Gimarc and Khan, *J. Am. Chem. Soc.* **100**, 2340 (1978). Copyright by the American Chemical Society.]

The angular change from the regular tetrahedron (T_d) to the diagonally folded square (C_{2v}) completely removes the degeneracy within the t_1 sets, each set yielding discrete orbitals classified as a_1, b_1, and b_2 under C_{2v} symmetry. The energy of $b_2(C_{2v})$ is the same as that of its $t_1(T_d)$ relative because the ligands that move remain on the b_2 MO nodal surface where the AOs of the moving ligands can make no contribution to the MO. Since no overlap changes occur in $t_1(T_d)$-$b_2(C_{2v})$, there is no energy change. The energy of $1b_1(C_{2v})$ falls below that of $1t_1(T_d)$ because the ligand orbitals move into the most favorable overlapping positions on the axis of the central atom p orbital. The $2b_1$ orbital rises in energy because the angular change increases the out-of-phase overlap between ligand and central atom p AOs. As the folded square flattens out to become planar (D_{4h}), the orbitals b_1 and $b_2(C_{2v})$ become degenerate as $e_u(D_{4h})$, as one can easily see from their AO composition diagrams. Through this angular change the energies of the b_1 orbitals remain constant because the moving ligands lie on nodal surfaces of the b_1 orbitals. The energy of $2a_1(C_{2v})$ increases relative to $1b_1(T_d)$ because AOs move away from positions of good in-phase overlap with the central atom p and toward the nodal surface of the MO. The energy of antibonding $4a_1(C_{2v})$ orbital decreases because similar but out-of-phase overlaps are relaxed. The contributions of AOs on the axial ligands in $2a_1$ and $4a_1(C_{2v})$ need not be zero, but from these arguments alone one would expect them to be small and, hence, they are omitted in Fig. 2.

Consider the $2a_1(T_d)$-$3a_1(C_{2v})$ orbital. Opening a tetrahedral angle produces no overlap changes between the ligand AOs and the central atom s AO, but there are overlap changes among the ligand orbitals themselves. Although these overlaps are rather small because of the relatively long ligand-ligand distances, the overlap changes are amplified by the fact that the ligand AO coefficients in $2a_1(T_d)$ are large due to the large number of phase differences within that MO. Opening a tetrahedral angle to diagonally folded square geometry moves ligands closer together, increasing the in-phase overlaps among ligand orbitals and lowering the energy of $3a_1(C_{2v})$ relative to that of $2a_1(T_d)$. In tetrahedral geometry $2a_1(T_d)$ has six ligand-ligand overlaps that are $109\frac{1}{2}°$ apart. In folded square geometry $3a_1(C_{2v})$ has four ligand-ligand overlaps that have closed to $90°$ apart, one set still at $109\frac{1}{2}°$, and one set that has opened to $180°$. Moving on from diagonally folded square to square planar (D_{4h}) shape, only one angle changes and one ligand-ligand overlap decreases when its angle opens from $109\frac{1}{2}°$ to $180°$. The other five angles and overlaps remain constant. Therefore, the energy should increase from $3a_1(C_{2v})$ to $2a_{1g}(D_{4h})$ and this intended correlation is indicated by a dashed connecting line in Fig. 2.

The orbitals $1t_1$-$2a_1$-a_{2u} and $2t_1$-$4a_1$-b_{1g} require special examination. As the tetrahedron flattens to the plane, the ligand orbitals move from good overlap positions in $1t_1(T_d)$ to the nodal plane of the central atom p orbital in $a_{2u}(D_{4h})$. Thus, in square planar geometry the ligand AOs all drop out of the MO, leaving $a_{2u}(D_{4h})$ as a free central atom p or nonbonding MO. The same process might appear to proceed in the higher energy $2t_1$-$4a_1$-b_{1g} case. Here, out-of-phase interactions are reduced as ligand AOs move away from the lobes of the central atom p orbital and toward the nodal surface of the central atom p, as in $4a_1(C_{2v})$. However, further flattening cannot cause the ligand AO contributions to vanish from this MO because that would reproduce the $a_{2u}(D_{4h})$ orbital we already have at lower energy. Instead, the central atom p vanishes leaving $b_{1g}(D_{4h})$ as four ligand AOs of alternate phase, just as they were in the $2t_1$ component from which this MO originated. A dashed tie line in Fig. 2 shows the intended correlation of $4a_1(C_{2v})$ and $b_{1g}(D_{4h})$. On the energy scale we have placed the nonbonding b_{1g} orbital below the antibonding $2a_{1g}(D_{4h})$ MO, although with a small central atom and large ligands, ligand-ligand out-of-phase overlaps might make b_{1g} antibonding.

We assumed we were studying a system with a rather electronegative central atom. Therefore, we feel justified in placing the central atom pure p nonbonding a_{2u} MO below the nonbonding b_{1g} MO that contains ligand-ligand antibonding interactions. Notice that b_{1g} and $2a_{1g}(D_{4h})$ would both be classified as a_1 under C_{2v} symmetry. Since orbitals of the same symmetry cannot cross, the intended correlations $3a_1(C_{2v})$-$2a_{1g}(D_{4h})$ and $4a_1(C_{2v})$-$b_{1g}(D_{4h})$ are not allowed. Instead, the actual connections are those shown with solid lines in Fig. 2.

The noncrossing of intended correlations between C_{2v} and D_{4h} MOs does introduce some uncertainty about where $2a_{1g}(D_{4h})$ lies on the energy ladder relative to $4a_1(C_{2v})$ and how $b_{1g}(D_{4h})$ compares with $3a_1(C_{2v})$. Overlap arguments put $2a_{1g}(D_{4h})$ below $2a_1(T_d)$ and, therefore, probably well below $4a_1(C_{2v})$ also. Overlap arguments also place $3a_1(C_{2v})$ below $2a_{1g}(D_{4h})$, and therefore, the antibonding $3a_1(C_{2v})$ orbital might have an energy that is comparable to or not much higher than nonbonding $b_{1g}(D_{4h})$. Since $2a_1(T_d)$-$3a_1(C_{2v})$-$b_{1g}(D_{4h})$ is the highest occupied MO in ten-electron AB$_4$ complexes, the size of the energy difference between $3a_1(C_{2v})$ and $b_{1g}(D_{4h})$ is crucial to structural conclusions for these complexes. If $3a_1$ is too high, the energy drop from $3a_1$ to b_{1g} will overcome the energy rise of the lower occupied $2a_1(C_{2v})$-$a_{2u}(D_{4h})$ system and the model will predict that ten-electron AB$_4$ complexes should be square planar, contrary to observation. This difficulty has troubled previous similar models. In our model we have assumed central atom and ligand relative AO energies as shown in Fig. 1a, placing $2a_1$ below $2t_1$ for tetrahedral geometry. This guarantees that $3a_1(C_{2v})$ will be even lower in energy than $2a_1(T_d)$ and, in particular, lower than the $4a_1(C_{2v})$ orbital from which $b_{1g}(D_{4h})$ originates.

Orbital mixing provides still another argument for a relatively low energy for $3a_1(C_{2v})$ and it gives more realistic representations of the C_{2v} MOs. Symmetry does not require the absence of a horizontal p orbital on the central atom in $3a_1(C_{2v})$ as it does for the related orbitals $2a_1(T_d)$ and $2a_{1g}(D_{4h})$. Similarly, a central atom s orbital could enter $4a_1(C_{2v})$ but not the related $2t_1(T_d)$ component or $b_{1g}(D_{4h})$. Mixing the two highest energy orbitals of a_1 classification stabilizes $3a_1$ and destabilizes $4a_1(C_{2v})$. After mixing, the $3a_1$ MO, with a large lobe pointing electron density away from the vertex of the two equatorial ligand bonds, looks much like the nonbonding lone pair orbital of the VSEPR model. Mixing stabilizes $3a_1$ (Fig. 3).

The correlation diagram of Fig. 2 can now be considered as a whole. For eight-electron complexes the bonding $1t_1(T_d)$ MOs are fully occupied and higher energy orbitals are empty. Angular changes to C_{2v} or D_{4h} shapes lead to a considerable total energy increase; therefore, the eight-electron complexes are tetrahedral. For ten-electron complexes the highest occupied orbital system is $2a_1(T_d)$-$3a_1(C_{2v})$-$b_{1g}(D_{4h})$. The large energy drop from

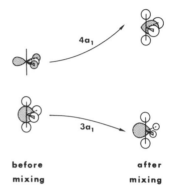

before
mixing

after
mixing

$2a_1(T_d)$ to $3a_1(C_{2v})$ outweighs the increase from $1t_1(T_d)$ to $2a_1(C_{2v})$ below and gives the ten-electron complexes the diagonally folded square structure. The drop must be steep enough to give the nine-electron radicals the C_{2v} structure as well. We have argued that the slope downward from $3a_1(C_{2v})$ to $1b_{1g}(D_{4h})$ is a modest one. Therefore, at least by itself, it is not enough to overcome the rising $2a_1(C_{2v})$-$a_{2u}(D_{4h})$ level below and ten-electron complexes are held at the diagonally folded square structure. The square planar structure must not be too much higher in energy, however. The fact that SF$_4$ undergoes axial-equatorial ligand exchange at room temperatures (*80*) means that square planar geometry must be only a few kcal per mole higher than the folded square structure. Finally, in twelve-electron complexes the orbital system $2t_1(T_d)$-$4a_1(C_{2v})$-$2a_{1g}(D_{4h})$ is occupied. The energy drop across this system makes 11- and 12-electron complexes rigidly square planar.

At this point we should go back to see what would have happened had we constructed the T_d-C_{2v}-D_{4h} correlation diagram assuming the energy-level ordering for tetrahedral geometry shown in Fig. 1b: $2t_1 < 2a_1$. Figure 4 shows the energy levels only. The $2a_1(T_d)$-$4a_1(C_{2v})$-$2a_{1g}(D_{4h})$ orbital system stretches across the top of the diagram. The energy-level systems flowing from $1t_1(T_d)$ and $2t_1(T_d)$ are practically mirror images of each other. The continuous and steep energy lowering from $2t_1(T_d)$ to $3a_1(C_{2v})$ to $b_{1g}(D_{4h})$ leads to square planar geometry for the ten-electron complexes. Indeed, the diagram shown in Fig. 4 may be appropriate for some systems. For example, in excited methane (eight electrons) one electron has been removed from the tetrahedral stabilizing $1t_1$-$2a_1$-a_{2u} system and added to the $2t_1$-$3a_1$-b_{1g} orbital that produces the square planar shape. The excited state of methane is believed to be square planar rather than the folded square (*93*).

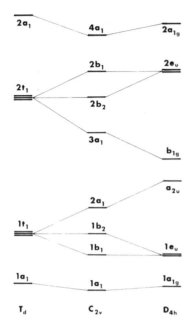

FIG. 4   MO correlations for T_d, C_{2v}, and D_{4h} symmetries assuming T_d levels from Fig. 1b.

Another structure that has been considered for four-coordinate complexes is the square pyramid (C_{4v}). Figure 5 correlates MOs of square pyramidal geometry with those for diagonally folded square and square planar shapes. There are three orbitals of a_1 symmetry for C_{4v} and we have already mixed the two of highest energy, $2a_1$ and $3a_1(C_{4v})$. They are very similar to the related C_{2v} orbitals shown in Fig. 3. The system $2a_1(C_{2v})$-$2a_1(C_{4v})$-$a_{2u}(D_{4h})$ is the highest occupied orbital for eight-electron complexes. For ten electrons, the highest occupied MO system is $3a_1(C_{2v})$-$b_1(C_{4v})$-$b_{1g}(D_{4h})$. The AO compositions of $b_1(C_{4v})$ and $3a_1(C_{2v})$ are quite different, but symmetry requires that they be connected since b_1 is classified as a_1 under C_{2v} symmetry. Clearly, C_{4v} geometry, with the close association of four ligand AOs of alternate phase in $b_1(C_{4v})$, is a high-energy structure for ten-electron complexes. For 12 electrons the $4a_1(C_{2v})$-$3a_1(C_{4v})$-$2a_{1g}(D_{4h})$ system is filled. The orbital $3a_1(C_{4v})$ is higher in energy than $2a_{1g}(D_{4h})$ because $3a_1$ has its ligand AOs pushing into a big central lobe of opposite phase. The energies of $3a_1(C_{4v})$ and $4a_1(C_{2v})$ may be comparable. In $4a_1(C_{2v})$ the equatorial ligand AOs are large contributors and they are very close to a large central lobe of opposite phase. As the axial ligands fold toward each other to form the square pyramid, the equatorial ligand AO coefficients shrink and the axial AO coefficients grow. In $3a_1(C_{4v})$ contributions from all four ligand AOs are equal. It is clear that the square pyramidal structure is high in energy compared to the folded square and square planar shapes.

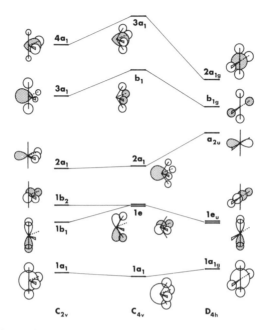

FIG. 5 MO correlations for AB₄ complexes in diagonally folded square (C_{2v}), square pyramidal (C_{4v}), and square planar shapes (D_{4h}). [Reprinted with permission from Gimarc and Khan, *J. Am. Chem. Soc.* **100**, 2340 (1978). Copyright by the American Chemical Society.]

The diagonally folded square (C_{2v}) can be converted into a trigonal pyramid (C_{3v}) by bending one of the folded square axial ligands into the plane of the central atom and the two equatorial ligands to form the base of a trigonal pyramid. The fixed axial ligand of the folded square becomes the apical ligand of the trigonal pyramid. The relative stabilities of these structures are determined by the behavior of a single orbital (Fig. 6): $3a_1$ rises in energy for the angular geometry change from C_{2v} to C_{3v}. Both $3a_1(C_{2v})$ and $3a_1(C_{3v})$ have large central atom lobes, formed by mixing with the $4a_1$ MO of each system. In $3a_1(C_{3v})$ the three basal ligands have substantial co-efficients and are at right angles to and out-of-phase with the central atom lobe, while in $3a_1(C_{2v})$ there are only two such close out-of-phase interactions. Fewer out-of-phase overlaps give lower energy to $3a_1(C_{2v})$ and allow ten-electron complexes to avoid the trigonal pyramidal shape.

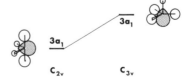

FIG. 6 Relative energies of the $3a_1$ MOs of AB₄ in folded square (C_{2v}) and trigonal pyramidal (C_{3v}) shapes.

OTHER PROPERTIES OF THE FOLDED SQUARE STRUCTURE

Reconsider for a moment the $3a_1$ MO of folded square geometry after mixing. Closing the equatorial angle to less than the tetrahedral value would increase overlaps in both $2a_1$ and $3a_1(C_{2v})$, leading to slightly lower energy. Table IV shows that the equatorial angles of SF_4 and SeF_4 are both less than $109\frac{1}{2}°$. Notice that orbital mixing has reduced the contributions from the equatorial ligand AOs in $3a_1$, reducing the electron density in the equatorial positions relative to that in the axial positions. Therefore, in complexes with mixed ligands such as XeO_2F_2 the more electronegative ligands should prefer axial positions where they can receive the extra electron density. This is in accord with the observation that the more electronegative ligands do occupy axial sites in mixed-ligand ten-electron complexes (80). The larger axial coefficients in $3a_1$ give a larger antibonding contribution to the $A\text{-}B_{ax}$ bond order than the smaller equatorial coefficients do to the $A\text{-}B_{eq}$ bond order, making the axial bonds longer and weaker than the equatorial bonds. The structural data in Table IV show that this is the case. Finally, if the axial ligands could lean away from the central atom lobe in $3a_1$, out-of-phase overlaps could be reduced. Again, the structural data in Table IV show that the axial ligands do lean slightly toward the equatorial ligands. [Apparently, the axial flourines in XeO_2F_2 lean away from the equatorial oxygens (86).]

Eight pairs of known isoatomic AB_4 complexes are:

$8e$: TlI_4^- $SnCl_4$ $PbCl_4$ $PbBr_4$ PbI_4 PBr_4^+ $AsCl_4^+$ $SbCl_4^+$
$10e$: TlI_4^{3-} $SnCl_4^{2-}$ $PbCl_4^{2-}$ $PbBr_4^{2-}$ PbI_4^{2-} PBr_4^- $AsCl_4^-$ $SbCl_4^-$

Each pair consists of an eight- and a ten-electron complex composed of the same number and kinds of atoms but with the central atoms differing by two in oxidation state. The additional electron pair in the ten-electron series

TABLE IV

Bond distances and angles for
ten-electron AB_4 and twelve-electron AB_6 complexes[a]

	SF_4	SF_6	SeF_4	SeF_6
$r(A\text{-}B_{ax})$ Å	1.646	—	1.771	—
$r(A\text{-}B_{eq})$ Å	1.545	—	1.682	—
$r(A\text{-}B)_{av}$ Å	1.595	1.564	1.727	1.688
$<B_{eq}\text{-}A\text{-}B_{eq}$	101.55°	—	100.55°	—
$<B_{ax}\text{-}A\text{-}B_{ax}$	173.07°	—	169.20°	—

[a] Reprinted with permission from Gimarc and Khan, *J. Am. Chem. Soc.* **100**, 2340 (1978). Copyright by the American Chemical Society.

occupies the antibonding $3a_1(C_{2v})$ MO. Therefore, the ten-electron complexes should have longer A-B bonds than the isoatomic eight-electron complexes. Unfortunately, no bond distances are known for the ten-electron complexes in this group.

LONE PAIR AND BOND PAIR ORBITALS FROM MOLECULAR ORBITALS

The lone pair and bond pair orbitals of conventional valence theory can be obtained from MO theory by taking linear combinations of MOs. AB$_4$ complexes in folded square geometry have five occupied MOs. One of these, the mixed $3a_1(C_{2v})$ MO with its large central atom lobe pointing away from the equatorial ligands, corresponds almost directly to the nonbonding lone pair orbital in the vacant position of the pseudotrigonal bipyramidal structure. The axial and equatorial ligand AO contributions can be trimmed away from the lone pair orbital by carefully mixing into $3a_1$ small amounts of $1a_1$, $2a_1$, and $1b_2$. Linear combinations of $1b_2$ and $2a_1(C_{2v})$ produce two equivalent bond orbitals for the equatorial ligands and combinations of $1b_1$, $1a_1$, and $3a_1$ can give the axial bond orbitals.

The six occupied MOs of square planar AB$_4$ complexes can produce the two lone pair and four equivalent bond pair orbitals of the localized or hybridized orbital picture. Suppose we wanted to use sp hybrids as the lone pair orbitals. A pure p orbital is already available as $a_{2u}(D_{4h})$ and we can regenerate an s orbital on the central atom by algebraically combining $1a_{1g}$ and $2a_{1g}$ MOs so that the ligand AOs cancel exactly. Thus, linear combinations of $1a_{1g}$, $2a_{1g}$, and a_{2u} MOs yield the two digonal or sp lone pair hybrids of conventional valence theory. Similarly, the four equivalent bond orbitals between central atom and ligands can be produced from four linear combinations of the $1a_{1g}$, $2a_{1g}$, b_{1g}, and the two $1e_u$ MOs.

Chapter 9 contains a more general discussion of the relationship between MOs and lone pair and bond pair orbitals.

STRUCTURES OF THE AB$_6$ COMPLEXES

Although a study of the AB$_5$ complexes might appear to be more systematic at this point, it is more logical and convenient to move on to the highly symmetric AB$_6$ complexes immediately because the MOs of the octahedral AB$_6$ class are closely related to those of the square planar AB$_4$ complexes we have just studied. Tables V and VI list the known AB$_6$ halides of main groups III through 0 of the periodic table (94–148). Table V contains complexes with 12 valence electrons and Table VI lists those with 14 valence electrons. Several 13-electron AB$_6$ radicals are also known. Examples are ClF_6, BrF_6, IF_6, SF_6^-, SeF_6^-, TeF_6^-, AsF_6^{2-}, and SbF_6^{2-} (149–151).

TABLE V

Known AB_6 halides, twelve valence electrons[a]

III	IV	V	VI	VII	0
AlF_6^{3-} (1.800)	SiF_6^{2-} (1.706)	PF_6^- (1.599) PCl_6^- (2.06) PBr_6^-	SF_6 (1.564)	ClF_6^+	
GaF_6^{3-} (1.808)	GeF_6^{2-} (1.77) $GeCl_6^{2-}$ (2.35)	AsF_6^- (1.67) $AsCl_6^-$	SeF_6 (1.688)	BrF_6^+	
InF_6^{3-} (2.04) $InCl_6^{3-}$ $InBr_6^{3-}$	SnF_6^{2-} (1.97) $SnCl_6^{2-}$ (2.42) $SnBr_6^{2-}$ (2.61) SnI_6^{2-} (2.85)	SbF_6^- (1.84) $SbCl_6^-$ (2.35) $SbBr_6^-$ (2.55)	TeF_6 (1.824)	IF_6^+	
TlF_6^{3-} (1.96) $TlCl_6^{3-}$ (2.49) $TlBr_6^{3-}$ (2.59)	PbF_6^{2-} (2.15) $PbCl_6^{2-}$ (2.50)	BiF_6^-			

[a] Reprinted with permission from Gimarc *et al. J. Am. Chem. Soc.* **100**, 2334 (1978). Copyright by the American Chemical Society.

TABLE VI

Known AB_6 halides, fourteen valence electrons[a]

III	IV	V	VI	VII	0
		PCl_6^{3-}		ClF_6^-	
				BrF_6^-	
		$AsCl_6^{3-}$	$SeCl_6^{2-}$ (2.40) $SeBr_6^{2-}$ (2.54) SeI_6^{2-}		
		SbF_6^{3-} $SbCl_6^{3-}$ (2.652) $SbBr_6^{3-}$ (2.799) SbI_6^{3-}	TeF_6^{2-} $TeCl_6^{2-}$ (2.541) $TeBr_6^{2-}$ (2.70) TeI_6^{2-} (2.90)	IF_6^-	XeF_6 (1.890)
	$SnBr_6^{4-}$ SnI_6^{4-}				
PbF_6^{4-} $PbCl_6^{4-}$ (2.93) $PbBr_6^{4-}$ (3.12) PbI_6^{4-}		BiF_6^{3-} $BiCl_6^{3-}$ (2.66) $BiBr_6^{3-}$ (2.840) BiI_6^{3-}	$PoCl_6^{2-}$ (2.54) $PoBr_6^{2-}$ (2.64) PoI_6^{2-} (2.82)		

[a] Reprinted with permission from Gimarc *et al. J. Am. Chem. Soc.* **100**, 2334 (1978). Copyright by the American Chemical Society.

Oxides with 12 valence electrons are SbO_6^{7-}, TeO_6^{6-}, IO_6^{5-}, and XeO_6^{4-} *(152–155)*. They seem to follow the same trend as the more numerous halides.

All AB_6 complexes for which structures have been determined are octahedral or nearly so. Why does nature seem to prefer the octahedral shape rather than some other reasonable (symmetric) conformation such as trigonal prismatic? Might other shapes be preferred by complexes that are as yet unknown?

The octahedron (O_h) can be converted into a trigonal prism (D_{3h}) by the rotation of one end of the structure by 60° relative to the other, assuming that A-B bond distances remain constant **(2)**. This internal rotation is

$$O_h \quad \text{⤚⤚⤚} \quad D_{3h}$$

2

related to that connecting staggered and eclipsed ethane. The particular orientation of structures emphasizes the comparison with the internal rotation in ethane.

Figure 7 is a correlation diagram for the rotation from the octahedral or staggered structure to the trigonal prismatic or eclipsed conformation.

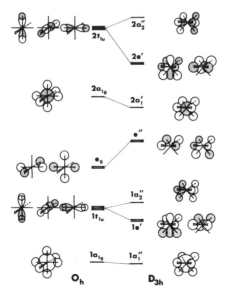

FIG. 7   MO correlations for octahedral (O_h) and trigonal prismatic (D_{3h}) structures for AB_6 complexes. [Reprinted with permission from Gimarc *et al.*, *J. Am. Chem. Soc.* **100**, 2334 (1978). Copyright by the American Chemical Society.]

We chose the orientations of structures in Fig. 7 for the ease of conceptual construction of the MOs. Octahedral symmetry allows doubly and triply degenerate orbitals. The triply degenerate levels must involve the three individual p AOs of the central atom each overlapping two of the six ligand orbitals. Because both in-phase and out-of-phase overlap combinations are possible there are two sets of the triply degenerate (t_{1u}) orbitals or a total of six MOs. Furthermore, there are two completely symmetric MOs (a_{1g}) made by in-phase or out-of-phase combinations of the central atom s orbital and the ligand AOs. The two remaining octahedral MOs must be the doubly degenerate e_g pair, represented in **3** as **a** and **b**. That these two

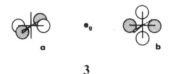

3

MOs have the same energy is obvious from their AO composition diagrams. Although **a** and **b** bring the total number of MOs to the required ten, one might be puzzled why the combination **c** (**4**) was not also included with the same energy as **a** and **b**. However, **c** is simply the algebraic sum of **a** and **b**. Since it is not an independent orbital **c** has not been included in our set. For degenerate orbitals it is possible to draw an infinite number of alternative

4

but equivalent AO composition diagrams. We chose to represent **a** and **b** in Fig. 7 for convenience only. In other circumstances it will be convenient to choose other representations.

The e_g MOs are nonbonding. In our model they contain no AO contributions from the central atom although they do have the symmetry of d AOs, and had we included central atom d orbitals in our basis set they would have appeared in e_g. The ligand AOs in e_g are out-of-phase with their cis neighbors.

The energy ordering of the O_h energy levels is straightforward. At lowest energy is the bonding $1a_{1g}$ orbital and above that the bonding $1t_{1u}$. The nonbonding e_g orbitals are at intermediate energy and above them are the antibonding $2a_{1g}$ and the antibonding $2t_{1u}$. The order and spacing of the energy levels in Fig. 7 are those obtained from extended Hückel calculations for model AH_6 systems.

Consider the rotation of AB_6 from staggered or octahedral to eclipsed or trigonal prismatic. The eclipsed a_1' orbitals should have lower energy than their staggered relatives a_{1g} because the ligand orbitals can overlap each other better in the eclipsed conformation in which they are closer together. In the staggered conformation the ligands are as far apart as possible. The bonding and nodeless $1a_{1g}$ and $1a_1'$ MOs have very nearly equal energies, but the higher energy antibonding set shows the predicted trend $a_{1g} > a_1'$. The t_{1u} orbitals split into an e' pair and a_2''. The energy decreases from $1t_{1u}$ (O_h) to $1e'$ (D_{3h}) (5). In the rotation keep the three ligands at the

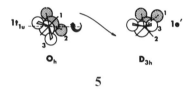

5

left-hand side of the O_h structure fixed and rotate the three right-hand ligands around the horizontal axis. Imagine the ligand 1 on the nodal surface in the O_h picture rotating back and away from the nodal surface that eliminates its AO from the O_h MO and into a position in D_{3h} where the ligand AO can overlap in-phase with the rear lobe of the central atom p AO. The ligand 2 that was to the rear in O_h rotates down and to the nodal surface where its AO vanishes in D_{3h}. The third rotating ligand 3, the one overlapping the front lobe of the central atom p in O_h, moves up a bit with no change in phase. The rotation moves in-phase ligand AOs closer together to increase overlap and lower the MO energy. Rotation raises the energy of the component of $1t_{1u}$ that becomes $1a_2''$ (6). Out-of-phase ligand AOs are closer together in

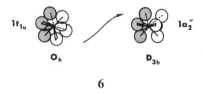

6

$1a_2''$ than they are in $1t_{1u}$. The energy increase should be small for at least two reasons: (1) The ligands are not particularly close to each other; that is, they are not bound to each other, and (2) although each rotating ligand AO moves closer to a ligand AO of opposite phase (its eclipsed partner), it also moves farther away from another ligand of opposite phase (one of its nearest neighbors in staggered geometry). Therefore, the energy increase of $1a_2''$

above $1t_{1u}$ is considerably smaller than the energy lowering of e' below $1t_{1u}$ and the extended Hückel results support the qualitative reasoning. Furthermore, since $1a_2'' > 1t_{1u}$ and $1e' < 1t_{1u}$, we can see that $1a_2'' > 1e'$ for the eclipsed or trigonal prismatic energy levels. Next, look at the energy increase of the octahedral or staggered e_g levels as they become the e'' levels of the trigonal prism or eclipsed conformation (7). The out-of-phase overlaps

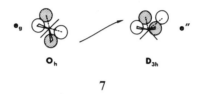

7

increase between ligands at opposite ends of the rotational axis producing an energy increase. These MOs are double-noded and the coefficients of the ligand AOs are large; therefore, the resulting energy changes occurring on rotation are large compared to those for $1a_2''$, $1e'$, or $1t_{1u}$ sets of lower energy.

Consider the correlation diagram of Fig. 7 as a whole. A 12-electron molecule such as SF_6 should be octahedral (staggered) because of the greater stability of the highest occupied orbitals e_g (O) relative to e'' (D_{3h}). The 12-electron AB_6 complexes are structurally rigid, as is the 12-electron A_2H_6 analog, diborane (to be studied in Chapter 6). Although a ten-electron complex would not have a closed shell of electrons in either conformation, we expect that the octahedral configuration would still be preferred. An eight-electron complex such as the hypothetical BeF_6 should be trigonal prismatic rather than octahedral. XeO_6 is another candidate for the trigonal prismatic structure. XeO_6^{4-} is known and octahedral. The electron affinity of the neutral XeO_6 might be too high to allow it to exist because of the empty e_g-e'' MOs.

Ab initio calculations of the floating spherical Gaussian type indicate that LiH_6^+ might be weakly bound (156). The ion has been detected experimentally (157). If the complex has a central atom, lithium is the logical choice and our model predicts that LiH_6^+ would be trigonal prismatic (D_{3h}) rather than octahedral (O_h). Ab initio calculations suggest that the six-electron ion H_7^+ has a structure that can be described as an H_3^+ ion with two H_2 units attached at two corners of the H_3^+ triangle and perpendicular to the H_3^+ plane (158). In other words, H_7^+ does not have an atom at the center of the structure. This is not unreasonable. A centrosymmetric structure such as D_{3h} requires a p AO on the central atom to make the occupied $1e'$ MOs bonding. The hydrogen $2p$ is high in energy above the $1s$ and, therefore, not likely to contribute to the $1e'$ MOs. On the other hand, the vacant $2p$ AOs of lithium are quite close to the $2s$ and, hence, much more likely to mix into the e' pair and make them bonding.

The 14-valence electron complexes such as XeF_6 and $IF_6{}^-$ would be isoelectronic with ethane and would be staggered (octahedral) rather than eclipsed (trigonal prismatic) for the same reasons that ethane is staggered (to be studied in Chapter 6): the antibonding dominant nature of the e_g and e'' orbitals, which lie below the highest occupied $2a_{1g}$-$2a_1'$ system (159). In the AB_6 case, however, the out-of-phase or antibonding interactions should be larger than those in the comparable A_2B_6 system because the end-end ligands are closer together in AB_6, making the rotational barrier for 14-electron AB_6 molecules somewhat larger than for A_2B_6.

The highest occupied MO in the 14-electron AB_6 system is the antibonding $2a_{1g}$ orbital. Although this orbital is completely symmetric with respect to all symmetry operations of the O_h group, this does not mean that the orbital is at a relative energy minimum with respect to angular variations. In fact, just the opposite is true; octahedral geometry represents an energy maximum for a_{1g} orbitals because the in-phase overlapping ligand orbitals are as far apart as they can possibly be. Any angular change would bring two or more ligand AOs closer together, increasing their overlaps and stabilizing the energy of the distorted MO. This must represent the driving force for distortions from pure octahedral symmetry. Several such distortions have been suggested to describe the fluxional behavior of XeF_6 (148,160). The experimental structural data for XeF_6 have recently been re-examined (161). Although XeF_6 easily distorts, isoelectronic species such as $TeCl_6^{2-}$, $TeBr_6^{2-}$, and $SbBr_6^{3-}$ seem to have stable regular octahedral structures. Our qualitative MO model can account for this trend. Two a_{1g} MOs can be made from the in-phase and out-of-phase combinations of the central atom s AO with the six ligand AOs. In a complex such as BiF_6^{3-}, one expects the ligand AO energies to lie far below that of the s AO on the central atom. If the energy gap between central atom s and ligand AOs is large (a large electronegativity difference between central atom and ligands), then the lower energy $1a_{1g}$ MO should be composed largely of ligand AOs while the $2a_{1g}$ MO should be mostly the central atom s AO with only small contributions from ligand AOs. If ligand AO coefficients are small in such a $2a_{1g}$ MO, a distortion from octahedral geometry will produce only a small energy lowering. See Fig. 8a. However, if the central atom-ligand AO energy gap is small, the ligand contributions to $2a_{1g}$ will be significant and a distortion from octahedral geometry will yield a large energy lowering. XeF_6 and Fig. 8b serve to illustrate this situation. Therefore, the trend should be an increased tendency to distortion from octahedral symmetry as the energy difference between central atom s and ligand AOs decreases. This is just what is observed.

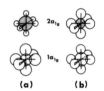

FIG. 8   Bonding and antibonding a_{1g} MOs of AB$_6$. (a) Central atom s AO is far above that of ligand AOs. (b) Ligand and central atom AOs are close in energy. [Reprinted with permission from Gimarc et $al.$, $J.$ $Am.$ $Chem.$ $Soc.$ **100**, 2334 (1978). Copyright by the American Chemical Society.]

BOND LENGTHS AND BOND STRENGTHS

Tables V and VI include some A-B bond distances. One might expect to see these distances increase from top to bottom of a group or column of the periodic table through a comparable series (e.g., SiF$_6^{2-}$, GeF$_6^{2-}$, SnF$_6^{2-}$, and PbF$_6^{2-}$, or SnF$_6^{2-}$, SnCl$_6^{2-}$, and SnI$_6^{2-}$) as atom size increases due to increasing principal quantum number and consequent increasing radius of the valence orbitals. The bond lengths should also decrease from left to right in a period as the same valence orbitals shrink due to increasing nuclear charge. Tables V and VI show that the bond lengths generally follow these trends. The deviations are at or near the level of uncertainties in the experimental data.

Several isoatomic AB$_6$ pairs are known, the members of which are composed of the same ligand and central atom but for which the oxidation state of the central atom differs by 2. The member with the higher oxidation state contains 12 valence electrons and the one with the lower oxidation state has 14. Table VII contains the pairs for which the A-B bond lengths are known for both members of the pair. In molecular orbital terms the difference between the members of an isoatomic pair is an electron pair occupying the antibonding $2a_{1g}$ MO of the 14-electron ion while $2a_{1g}$ is empty in the 12-electron member. An electron pair added to an antibonding MO should have the effect of weakening or lengthening the A-B bond, and the data in Table VII follow this rule. Other known isoatomic pairs include: ClF$_6^+$, ClF$_6^-$; BrF$_6^+$, BrF$_6^-$; IF$_6^+$, IF$_6^-$; and PbF$_6^{2-}$, PbF$_6^{4-}$. The member with the lower central atom oxidation state should have a considerably longer A-B bond length. Another measure of bond strengths is the symmetric stretching frequency $v_1(a_{1g})$, which is observed in the Raman spectrum.

TABLE VII

A-B bond distances (Å) for AB$_6$ pairs differing only in the oxidation state of the central atom[a]

| PbCl$_6^{2-}$ | 2.50 | SbCl$_6^-$ | 2.35 | SbBr$_6^-$ | 2.55 |
| PbCl$_6^{4-}$ | 2.93 | SbCl$_6^{3-}$ | 2.65 | SbBr$_6^{3-}$ | 2.80 |

[a] Reprinted with permission from Gimarc et $al.$, $J.$ $Am.$ $Chem.$ $Soc.$ **100**, 2334 (1978). Copyright by the American Chemical Society.

Again, the spectral data support our model: $PbCl_6^{2-}$ ($v_1 = 281$ cm^{-1}) (162); $PbCl_6^{4-}$ (202 cm^{-1}) (163); $SbCl_6^{-}$ (329 cm^{-1}) (162); $SbCl_6^{3-}$ (267 cm^{-1}) (163). In each case the 14-electron ion has the lower stretching frequency. This suggests that when comparing classes of molecules in which the MOs are filled by electrons to different levels, there is no reason to expect bond distances between different pairs of the same kinds of atoms to be equal as required by the simple notion of additivity of covalent radii. Where the additivity of radii works, it must be because comparable MO systems are occupied by electrons to the same level.

STRUCTURES OF AB₅ COMPLEXES

Table VIII lists the AB₅ complexes with ten valence electrons (164–178). These complexes are mainly trigonal bipyramidal (D_{3h}) in structure with the bonds to the axial ligands (A-B$_{ax}$) being slightly longer than those to the equatorial ligands (A-B$_{eq}$) (80) (8). Not all ten-electron AB₅ complexes

$$D_{3h} \qquad \qquad \qquad C_{4v}$$

8

have the D_{3h} structure in one or more phases. For example, although SbCl₅ is monomeric and trigonal bipyramidal in the gaseous, liquid and solid states (177b), PCl₅ is D_{3h} in the gas (173a) and in liquid solutions (173b) but in the solid it has the ionic form $PCl_4^+PCl_6^-$ (111). Solid PBr₅ is $PBr_4^+Br^-$ (174). In the crystalline state BiF₅ (178) and GeF₅⁻ (168b) consist of infinite chains of AB₅ units in which the central atom A is approximately octahedrally coordinated. More serious exceptions to trigonal bipyramidal geometry arise with $InCl_5^{2-}$ and $TlCl_5^{2-}$ (179,180). In the crystal, isolated $InCl_5^{2-}$ ions occur with square pyramidal geometry (C_{4v}). The bond to the apical chlorine (A-B$_{ap}$) is 2.415 Å and the somewhat longer basal chlorine bonds (A-B$_{ba}$) are 2.456 Å. The angle between apical and basal chlorines is 107.87° (179). The In atom is said to be above the basal plane of the square pyramid. $TlCl_5^{2-}$ is also square pyramidal.

The energy of the square pyramidal (C_{4v}) structure must not be far above that of the trigonal bipyramidal (D_{3h}) shape for the more characteristic members of the ten-electron series. NMR studies of PF₅, PCl₅ (181), SbCl₅, and SiF₅⁻ (182) indicate that all ligands are structurally equivalent on the

TABLE VIII

Known AB_5 complexes, ten valence electrons[a]

III	IV	V	VI	VII	0
AlF_5^{2-}	SiF_5^-	PF_5			
	$SiCl_5^-$	PCl_5			
		PBr_5			
GaF_5^{2-}	GeF_5^-	AsF_5			
	$GeCl_5^-$				
InF_5^{2-}		SbF_5			
$InCl_5^{2-}$	$SnCl_5^-$	$SbCl_5$			
	$SnBr_5^-$				
		BiF_5			
$TlCl_5^{2-}$					

[a] Reprinted with permission from Gimarc, *J, Am. Chem. Soc.* **100**, 2346 (1978). Copyright by the American Chemical Society.

NMR time scale despite the fact that axial and equatorial ligands are known from other structural studies to occupy positions with different electronic environments. Berry proposed a pseudorotation mechanism of intra-molecular ligand exchange to account for the observed NMR equivalence (*181*). In diagram 9, the axial ligands 4 and 5 on the left become equatorial on the right while equatorial ligands 2 and 3 become axial. The energy

9

barriers to pseudorotation have been estimated to be on the order of a few kcal per mole. Other mechanisms for intramolecular ligand exchange have been suggested including a process called the turnstile mechanism (*183*) (**10**). Here the trigonal bipyramid is assumed to distort to a structure of C_s symmetry, then ligands 2, 3, and 4 rotate by 60° about a pseudo-3-fold axis, like a turnstile, to another C_s structure that can relax to the bypyramidal shape with resulting exchange of axial and equatorial ligands.

10

A number of mixed ligand complexes are known in the ten-electron bipyramidal AB_5 series. These include PF_nCl_{5-n}, ClO_2F_3, IO_2F_3, and SOF_4 (184–187). The most stable configuration in each case has the more electronegative ligands in the axial positions (80).

The 12-electron AB_5 complexes listed in Table IX are square pyramidal (C_{4v}) (188–206). The bond to the apical ligand in each of these complexes is 0.1 to 0.2 Å shorter than the bonds to the basal ligands. The angle between apical and basal ligands is less than 90°, usually around 80°, and the central atom is said to be below the basal plane of the square pyramid, unlike that of the ten-electron square pyramidal structure of $InCl_5^{2-}$. Apical-basal intramolecular ligand exchange has not been observed in the 12-electron AB_5 complexes.

TABLE IX

Known AB_5 complexes, twelve valence electrons[a]

III	IV	V	VI	VII	0
			SF_5^-	ClF_5	
			SeF_5^- $SeCl_5^-$ $SeBr_5^-$	BrF_5	
		SbF_5^{2-} $SbCl_5^{2-}$ $SbBr_5^{2-}$ SbI_5^{2-}	TeF_5^- $TeCl_5^-$ $TeBr_5^-$	IF_5	XeF_5^+
	$PbBr_5^{3-}$	$BiCl_5^{2-}$ $BiBr_5^{2-}$ BiI_5^{2-}	PoI_5^-		

[a] Reprinted with permission from Gimarc, *J. Am. Chem. Soc.* **100**, 2346 (1978). Copyright by the American Chemical Society.

Examples of mixed ligand complexes of the 12-electron series are $TeOF_4^{2-}$, $ClOF_4^-$, $BrOF_4^-$, IOF_4^-, and $XeOF_4$ (207–211). In these complexes the apical position is occupied by the less electronegative ligand.

The 11-electron radicals SF_5, PF_5^-, and PCl_5^- are known (212–214). These are believed to have square pyramidal C_{4v} structures.

The conclusions of a number of ab initio and semiempirical calculations are remarkably consistent (215–226). The 12-electron ClF_5 complex is strongly square pyramidal (C_{4v}), not trigonal bipyramidal (D_{3h}). The ten-electron AB_5 systems prefer D_{3h} to C_{4v} geometry but only by a few kcal/mole. Calculated bond orders to axial positions in the D_{3h} structure are smaller than those to equatorial positions and, where geometries were obtained by total energy minimizations, the axial bonds are longer than the equatorial bonds. The calculated charge densities at the axial positions are more negative than those at the equatorial positions, indicating preferred axial sites for substitution of electronegative ligands. When Berry pseudorotation and turnstile mechanisms of intramolecular ligand exchange are compared for the phosphoranes or their analogs, the C_{4v} transition state of the pseudo-rotation mechanism turns out to be several kcal/mole lower than C_s structures for the turnstile mechanism. Finally, all of these conclusions appear to be qualitatively insensitive to whether or not central atom d AOs were included in the basis set.

MOLECULAR ORBITALS AND MOLECULAR SHAPES

AO composition diagrams of the MOs of D_{3h} geometry are shown in Fig. 9. The lowest energy MO is the bonding orbital $1a_1'$ composed of the central atom s and the ligand AOs all overlapping in-phase. The corresponding antibonding MO is $3a_1'$, with all five ligand AOs of opposite phase to that of the central atom s AO making the $3a_1'$ orbital high in energy. In $1a_2''$ the central atom p AO that points along the 3-fold axis overlaps in-phase the axial ligand AOs. The corresponding antibonding $2a_2''$ MO has the highest energy in the whole set. A system with a 3-fold symmetry axis must have doubly degenerate MOs. The lower energy degenerate set $1e'$ is composed of the central atom p AOs that lie in the equatorial plane overlapping in-phase with the equatorial ligand AOs. The energy of $1e'$ is higher than that of $1a_2''$ because the overlap between the ligand AOs and the central atom p orbital is poorer in $1e'$ than it is in $1a_2''$. The antibonding $2e'$ degenerate pair lies high in energy but below $2a_2''$.

Thus far we have formed eight of the nine MOs of the $AB_5(D_{3h})$ set. The last MO turns out to have a_1' symmetry. One way to derive this orbital is by considering the symmetry adapted basis functions of a_1' symmetry. Symmetry adapted basis functions have the desired symmetry, in this case

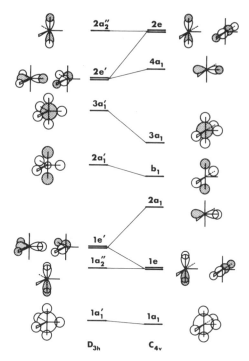

FIG. 9   MO correlations for AB$_5$ complexes in trigonal bypyramidal (D_{3h}) and square pyramidal (C_{4v}) shapes. [Reprinted with permission from Gimarc, *J. Am. Chem. Soc.* **100**, 2346 (1978). Copyright by the American Chemical Society.]

a_1', and they are composed of AO combinations in which only those AOs appear that have coefficients of equal magnitude as required by the symmetry of the group. Once in hand, the symmetry adapted basis functions can be linearly combined or mixed to form MOs. There are only three symmetry adapted functions of a_1' symmetry: $\mathbf{h}_1$, $\mathbf{h}_2$, and $\mathbf{h}_3$ (**11**). In $\mathbf{h}_1$ only the central atom s AO appears, $\mathbf{h}_2$ contains the two axial ligand AOs, and $\mathbf{h}_3$ is the sum of the three equatorial ligand AOs. From the three a_1' symmetry

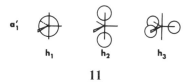

11

adapted basis functions, linear combinations can produce only three linearly independent a_1' MOs. We already have two of these, the bonding and antibonding $1a_1'$ and $3a_1'$ combinations, so the third is yet to be formed by putting together $\mathbf{h}_1$, $\mathbf{h}_2$, and $\mathbf{h}_3$ somehow. In both $1a_1'$ and $3a_1'$ we have combined $\mathbf{h}_2$ and $\mathbf{h}_3$ in-phase with each other, then we mixed that combination either in-phase with $\mathbf{h}_1$ to form $1a_1' = \mathbf{h}_1 + \mathbf{h}_2 + \mathbf{h}_3$, or out-of-phase with $\mathbf{h}_1$ to make $3a_1' = \mathbf{h}_2 + \mathbf{h}_3 - \mathbf{h}_1$. (In the MO combinations the symmetry adapted basis functions do not appear with the same weight, but since we are primarily interested in phase relations we ignore the mixing coefficients that should appear before $\mathbf{h}_1$, $\mathbf{h}_2$, and $\mathbf{h}_3$.) We already have two MOs with combination $\mathbf{h}_2 + \mathbf{h}_3$ so the third a_1' MO must have $\mathbf{h}_2$ and $\mathbf{h}_3$ of opposite phase: $\mathbf{h}_3 - \mathbf{h}_2$. But to the out-of-phase combination $\mathbf{h}_3 - \mathbf{h}_2$ we can match $\mathbf{h}_1$ either in-phase with $\mathbf{h}_3$ or in-phase with $\mathbf{h}_2$. Thus, we get two possibilities: $\mathbf{d} = \mathbf{h}_3 - \mathbf{h}_2 + \mathbf{h}_1$ and $\mathbf{e} = \mathbf{h}_3 - \mathbf{h}_2 - \mathbf{h}_1$. Either of these combinations looks like the kind of MO that might lie in energy between $1a_1'$ and $3a_1'$. How do we choose between them? It would be tempting to combine $\mathbf{d}$ and $\mathbf{e}$ to make $\mathbf{f}$, canceling out the central atom s AO as in **12** (*215,216*). The choice of $\mathbf{f}$ is appealing and, in fact, calculations yield a small

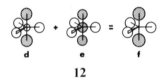

12

central atom s AO coefficient, but that selection would discard some very useful information about $2a_1'$ that makes it different from the related b_1 MO of C_{4v} geometry. D_{3h} symmetry allows the central atom s AO to enter a_1' MOs and even though its contribution to $2a_1'$ may be small, it will directly influence the energy of that MO. On the other hand, the central atom s coefficient of the related b_1 MO of C_{4v} symmetry must be zero.

The relative energies of the a_1' MOs are related to the numbers of A-B bonding and antibonding interactions. For example, $1a_1'$ has five A-B bonding interactions and is low in energy while $3a_1'$ has five A-B antibonding overlaps and is high in energy. The orbital $2a_1'$ (**d**) has three A-B bonding overlaps involving the equatorial ligands and two A-B antibonding overlaps with the axial ligands. Therefore, $2a_1'$ (**d**) must have lower energy than the $2a_1'$ (**e**) picture that has two A-B bonding and three A-B antibonding interactions. Figure 9 contains the picture $2a_1'$ (**d**) because of its lower energy and its explicit central atom s AO content.

The C_{4v} structure is obtained from the D_{3h} shape by opening the angle between a pair of equatorial ligands from 120 to 180°. This makes the apical-basal angle 90°, which is greater than that for the 12-electron complexes such as IF_5 but less than that for the 10-electron complex $InCl_5^{2-}$. The 90° angle assumed in Fig. 9 simplifies energy comparisons and serves as a compromise C_{4v} structure. The MOs of C_{4v} symmetry develop from those of D_{3h} geometry.

Angular changes that increase AO overlap lower the MO energy. The energies of the two $a_2''(D_{3h})$-$e(C_{4v})$ MOs are the same because their AO compositions and overlaps are identical. The energy of $1e(C_{4v})$ is lower than that of $1e'(D_{3h})$ because overlap between the central atom p and ligand AOs is better in $1e(C_{4v})$. The MO energy increases from $1e'(D_{3h})$ to $2a_1(C_{4v})$ because a pair of ligand AOs in good overlap with the central atom p AO in one of the components of $1e'(D_{3h})$ move away from the p-orbital lobe and towards the MO nodal surface in $2a_1(C_{4v})$. This is a large change in overlap, and therefore energy, compared to that of $1e'(D_{3h})$-$1e(C_{4v})$.

The $2a_1'(D_{3h})$-$b_1(C_{4v})$ orbital system requires special discussion. The central atom s and one of the equatorial ligand AOs drop out of $2a_1'(D_{3h})$ when it becomes $b_1(C_{4v})$ and the nodal surfaces move to cut through the base edges of the pyramid and intersect each other along the C_4 axis. The alternate phases of the basal ligand AOs of $b_1(C_{4v})$ are the same as those of the axial and equatorial ligand AOs in the $2a_1'(D_{3h})$ MO to which b_1 is related. Canceling A-B bonding and antibonding interactions should give $2a_1'$ (d) an energy in the nonbonding range and therefore comparable to $b_1(C_{4v})$, but whether $2a_1'(D_{3h})$ is above or below b_1 is uncertain from qualitative arguments alone. One might be more confident in saying that $2a_1'$ (e) is higher than b_1. Extended Hückel (216) and ab initio (215) MO calculations on AB_5 model systems put $2a_1'(D_{3h})$ slightly higher than $b_1(C_{4v})$. The point here is that the energy difference between $2a_1'$ and b_1 is not usually large enough to control the molecular shape of ten-electron AB_5 complexes for which $2a_1'$-b_1 is the highest occupied MO. Instead, geometry is primarily determined by the steeply rising $1e'(D_{3h})$-$2a_1(C_{4v})$ MO system just below.

The orbital correlation $3a_1'(D_{3h})$-$3a_1(C_{4v})$ also merits extra consideration. Opening the equatorial angle in $3a_1'(D_{3h})$ moves equatorial ligands closer together and increases the overlaps in two pairs of ligand AOs. Although the overlaps themselves are not large, the ligand AO coefficients in $3a_1'$-$3a_1$ are large because of the complicated nature of the nodal surfaces in the orbitals and the large coefficients amplify the overlap change to make $3a_1(C_{4v})$ considerably lower in energy than $3a_1'(D_{3h})$. However, orbital mixing also acts to stabilize $3a_1(C_{4v})$ relative to $3a_1'(D_{3h})$. Symmetry eliminates any contribution to $3a_1'(D_{3h})$ from the horizontal (equatorial) p AO of the central atom, but this AO could add to $3a_1(C_{4v})$. Similarly, no central atom s orbital can be included in $2e'(D_{3h})$ although it is allowed in $4a_1(C_{4v})$. The orbitals $3a_1$ and $4a_1(C_{4v})$ are the two highest energy MOs of a_1 classification and their AO compositions are different. Therefore, they should be mixed, as shown in Fig. 10. Mixing makes $3a_1$ an orbital with a large central atom lobe pointing out of the base of the square pyramid and of phase opposite to that of the four basal ligand AOs. The axial ligand AO contribution is reduced but not necessarily to zero as assumed above. The

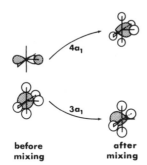

FIG. 10   Before mixing and after mixing representations of $3a_1$ and $4a_1$ MOs of AB$_5$ in square pyramidal (C_{4v}) geometry. [Reprinted with permission from Gimarc, *J. Am. Chem. Soc.* **100**, 2346 (1978). Copyright by the American Chemical Society.]

before mixing   after mixing

perturbation effect of mixing is to lower the energy of $3a_1$ and raise that of $4a_1$. The aftermixing picture of $3a_1(C_{4v})$ reminds one of the lone pair orbital of the valence shell electron pair repulsion (VSEPR) model. Orbital mixing and increased ligand-ligand AO overlaps make the energy of $3a_1(C_{4v})$ considerably lower than that of $3a_1'(D_{3h})$. Mixing also accounts for the higher than expected energy of $4a_1(C_{4v})$, as shown in Fig. 9. On overlap arguments alone one would have expected $4a_1(C_{4v})$ to be lower than $2e'(D_{3h})$.

Consider the correlation diagram of Fig. 9 as a whole. For ten-electron AB$_5$ complexes the highest occupied MO system is $2a_1'(D_{3h})$-$b_1(C_{4v})$. Although the energy slope of this system may favor the square pyramidal structure, the energy difference is not enough to overcome the steeply rising $1e'(D_{3h})$-$2a_1(C_{4v})$ orbital of lower energy. It is the $1e'$-$2a_1$ system that gives the ten-electron AB$_5$ complexes their trigonal bipyramidal geometry. The 12-electron complexes are square pyramidal because the $3a_1'(D_{3h})$-$3a_1(C_{4v})$ MO system is occupied.

THE BARRIER TO PSEUDOROTATION

Ligand exchange has been observed experimentally in a number of ten-electron AB$_5$ complexes. If this occurs by the Berry pseudorotation mechanism, and the best evidence indicates that it does, it is because the C_{4v} transition state is only a few kcal per mole above the D_{3h} structure for these complexes. The source of the pseudorotation barrier is the $1e'(D_{3h})$-$2a_1(C_{4v})$ MO in Fig. 9, but this may be reduced or damped by the $2a_1'(D_{3h})$-$b_1(C_{4v})$ system above it. Now $b_1(C_{4v})$ is the only MO of either symmetry that is composed of ligand AOs only and therefore independent of the energy of the central atom orbitals. The choice of a central atom of low electronegativity and therefore high-energy s AO will raise the energy of $2a_1'(D_{3h})$ but not $b_1(C_{4v})$. The choice of more electronegative ligands with lower energy AOs will lower $b_1(C_{4v})$ more than it will $2a_1'(D_{3h})$. Thus by appropriate choices of ligands and central atoms we can control the barrier to pseudorotation. The rules are these: (1) Lower central atom electronegativity lowers the barrier, and (2) greater ligand electronegativity lowers the barrier. Consider the series PF_5, PCl_5, and PBr_5. The ligand electronegativity decreases relative to a constant central atom through this series. Since large ligand electronegativity favors a low pseudorotation barrier, we predict that the barrier should increase as $PF_5 < PCl_5 < PBr_5$. Through the series PF_5, AsF_5, and SbF_5 the electronegativity of the central atom decreases relative to the constant ligand. Since lower central atom electronegativity lowers the barrier, the order of barriers should be $SbF_5 < AsF_5 < PF_5$. Through the series SbF_5, $AsCl_5$, and PBr_5 the central atom electronegativities increase and the ligand electronegativities decrease, both raising the barrier. Therefore, we predict that barriers should increase as $SbF_5 < AsCl_5 < PBr_5$. For the series PF_5, $AsCl_5$, and $SbBr_5$, central atom electronegativities decrease favoring lower barriers while ligand electronegativities also decrease but favoring higher barriers. Because the effects oppose each other, no qualitative predictions are possible in this case.

How do these predictions compare with estimated barriers from experiment? In his original interpretation of NMR data for PF_5 and PCl_5, Berry concluded that ligands exchange faster in PF_5 than in PCl_5 and therefore PF_5 must have a lower barrier than does PCl_5 or $PF_5 < PCl_5$, the order predicted by the qualitative model (181). A careful study of the vibrational spectra of PF_5 and AsF_5, including calculated potential functions for axial-equatorial exchange, shows that $AsF_5 < PF_5$ (227), which checks the order predicted by the qualitative model. Several tables of estimated barriers are available based on vibrational spectral data and various assumptions (228). Unfortunately, there are a number of reversals of relative barrier size in those compilations but two consistencies remain: $SbCl_5 < PCl_5$ and $SbCl_5 < AsF_5$. The first pair matches the order that would be predicted by the qualitative model while the second pair presents a comparison the qualitative model cannot make because of opposing effects. In an ab initio SCF MO study of PH_5, an additional calculation was performed in which the nuclear charge on the hydrogens was increased from 1.0 to 1.1, making them slightly more electronegative (215). The effect was to lower slightly the barrier to pseudorotation compared to that for normal hydrogens. Once more, this is just what the qualitative model predicts would result from increasing the ligand electronegativity. Thus where more rigorous data are available for comparison, the qualitative MO predictions are correct.

Note that the pseudorotation phenomenon is one of ten-electron AB_5 complexes in general and not restricted to group V halides. For example, the qualitative model predicts the following increasing order of barriers: $AlF_5^{2-} < SiF_5^- < PF_5$. No experimental data are available for comparison.

Is it possible to make the energy of $2a_1'(D_{3h})$ so high and that of $b_1(C_{4v})$ so low that the pseudorotation barrier disappears entirely and the ten-electron complex is more stable in square pyramidal form? Among the elements considered here Tl has the highest energy valence s AO and combining ligands of lowest energy would yield TlF_5^{2-}. Unfortunately, this complex is not known but the closely related species $InCl_5^{2-}$ and $TlCl_5^{2-}$ both have square pyramidal shape. Other promising candidates for C_{4v} structures are PbF_5^- and $PbCl_5^-$, but these complexes are unknown.

NMR data indicate that the ten-electron complex SF_4 undergoes axial-equatorial ligand exchange through a Berry pseudorotation mechanism (229). Barriers must be on the order of a few kcal per mole, comparable in size to those of the ten-electron AB_5 series. Although ten-electron AB_4 and AB_5 systems are floppy, the eight-electron AB_3 systems are much more rigid. Figure 11 shows the higher occupied MOs of AB_3, AB_4, and AB_5 in order to compare similarities and differences. The comparison is one between complexes formed from the same central atoms and ligands such as PF_3, PF_4^-, and PF_5.

The origin of the inversion barrier is a strikingly similar MO in all three cases. It is the $2a_1(C_{3v})$-$1a_2''(D_{3h})$ MO system in AB_3, $2a_1(C_{2v})$-$a_{2u}(D_{4h})$ in AB_4, and $1e'(D_{3h})$-$2a_1(C_{4v})$ in AB_5. In each instance the barrier results from ligand AOs moving away from the central atom p AO that lies along the principal symmetry axis of the transition state. At the top of the barrier,

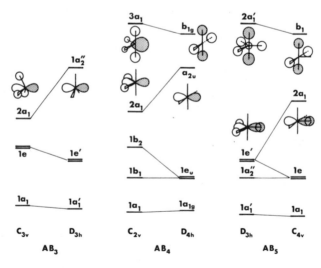

FIG. 11 Comparison of higher occupied MOs and barriers to inversion and pseudorotation for $AB_3(8e)$, $AB_4(10e)$, and $AB_5(10e)$ complexes. [Reprinted with permission from Gimarc, *J. Am. Chem. Soc*, **100**, 2346 (1978). Copyright by the American Chemical Society.]

two or more ligands are on or near the nodal surface of the MO. But in the AB_4 and AB_5 cases a higher occupied MO acts to damp or lower the barrier: $3a_1(C_{2v})$-$b_{1g}(D_{4h})$ in AB_4 and $2a_1'(D_{3h})$-$b_1(C_{4v})$ in AB_5. Even these damping MOs have much in common. Both $3a_1(AB_4, C_{2v})$ and $2a_1'(AB_5, D_{3h})$ are nonbonding MOs that include central atom AOs. At the transition state, the orbitals $b_{1g}(AB_4, D_{4h})$ and $b_1(AB_5, C_{4v})$ are identical and involve only ligand AOs. This suggests that the ten-electron AB_4 complexes should follow the same rules as AB_5 complexes for predicting relative heights of barriers to pseudorotation. No barriers are known for the AB_4 series. The qualitative MO model cannot predict whether AB_4 barriers should be higher or lower than those for AB_5, but it can explain why the AB_3 complexes, lacking the higher energy barrier-damping MO, should have higher inversion barriers than either AB_4 or AB_5.

BOND DISTANCES AND ANGLES AND SUBSTITUENTS

The data for ten-electron AB_5 complexes in Table X show that the axial A-B bonds are slightly longer than the equatorial A-B bonds. The differences are small but consistent. Bond order contributions from individual MOs are proportional to the overlap of AOs on connected atoms multiplied by the product of coefficients of the overlapping AOs. The nodeless MO $1a_1'(D_{3h})$ is smooth and spherical and ligand AO coefficients are practically equal. The $2a_1'$ orbital is nearly A-B nonbonding because of the small coefficient of the central atom s AO. Greater differences in bond order occur in the orbitals $1a_2''$ (A-B_{ax} bonding only) and $1e'$ (A-B_{eq} bonding only). The $1e_y'$ orbital (viewed in **13** down the 3-fold axis) has the same AO composition as $1a_2''$, but with each ligand tilted 30° away from the central atom p axis in $1e_y'$ the overlap is still 87% ($= \cos 30°$) of the maximum value that is possible in $1a_2''$ (100%). Therefore, one can expect the A-B_{eq} bond order contribution from $1e_y'$ to be about 87% of the A-B_{ax} contribution of $1a_2''$. But $1e_x'$ also contributes to the A-B_{eq} bond order. Considering the same

TABLE X

Bond distances (Å) for ten electron AB_5 complexes in the trigonal bipyramidal structure[a]

	$SnCl_5^-$	PF_5	PCl_5	AsF_5	$SbCl_5$
A-B_{ax}	2.38	1.577	2.124	1.711	2.34
A-B_{eq}	2.36	1.534	2.020	1.656	2.29

[a] Reprinted with permission from Gimarc, *J. Am. Chem. Soc.* **100**, 2346 (1978). Copyright by the American Chemical Society.

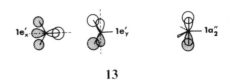

13

pair of ligands in $1e_y'$, the corresponding ligand AOs in $1e_x'$ are 60° from the central atom p axis, with an overlap that is 50% ($= \cos 60°$) of maximum. The $1e_x'$ orbital contains two ligands AOs of the same phase and, therefore, MO normalization requires that the coefficients of those two AOs be smaller in $1e_x'$ than in $1e_y'$. Still, $1e_x'$ provides a healthy contribution to the $A\text{-}B_{eq}$ bond order. The combined overlap fraction for the $1e'$ pair is 1.37 ($= 0.87 + 0.50$) compared to 1.00 for $1a_2''$, only slightly larger than the ratio of calculated A-B bond order contributions from $1e'$ and $1a_2''$ (1.25 for a model AH_5 system). Thus, $A\text{-}B_{eq}$ bonds are shorter than $A\text{-}B_{ax}$ bonds because the two $1e'$ orbitals combine to give a larger contribution to $A\text{-}B_{eq}$ than the single $1a_2''$ orbital produces for $A\text{-}B_{ax}$. A simple way to summarize this result is to say that the three equatorial ligands form bonds involving two central atom p AOs, while the two axial ligands must compete to bond through only one central atom p.

Calculated charge densities are larger at the axial sites than at the equatorial positions. This is a result of MO orthogonality. Suppose we orthogonalize $2a_1'$ (**d**) (**12**) against the spherical blob of $1a_1'$. The result requires larger ligand AO coefficients in $2a_1'$ (**d**) at the axial positions where there are only two AOs than at equatorial positions where three AOs appear. Larger coefficients at the axial sites put larger electron densities there. More electronegative substituents prefer positions where coefficients are larger and which offer greater electron density.

In each 12-electron complex the apical bond is shorter than the basal bonds. The data in Table XI show differences that range from 0.03 to 0.3 Å. The highest occupied MO is $3a_1(C_{4v})$. Orbital mixing reduces the apical

TABLE XI

Bond distances (Å) and angles (°) for 12 electron AB_5 complexes in square pyramidal geometry[a]

	SbF_5^{2-}	TeF_5^-	IF_5	XeF_5^+	BrF_5	$SbCl_5^{2-}$
$A\text{-}B_{ap}$	1.916	1.862	1.844	1.793	1.689	2.356
$A\text{-}B_{ba}$	2.075	1.953	1.869	1.845	1.774	2.636
$B_{ap}\text{-}A\text{-}B_{ba}$	79.4	78.1	81.9	79.0	84.8	85.2

[a] Reprinted with permission from Gimarc, *J. Am. Chem. Soc.* **100**, 2346 (1978). Copyright by the American Chemical Society.

ligand AO coefficient, reducing the A-B_{ap} antibonding nature of $3a_1$. Furthermore, the central atom p AO introduced into $3a_1$ by MO mixing enters in-phase with the apical ligand to produce a small bonding component in the A-B_{ap} bond order from $3a_1$. Thus, $3a_1$ is essentially A-B_{ap} nonbonding but still A-B_{ba} antibonding. The b_1 MO makes no contribution to either A-B_{ap} or A-B_{ba} bond orders because this orbital contains no central atom AOs. Compare $2a_1$ with the two $1e$ orbitals. The overlaps are exactly the same in the two different kinds of MOs, but there are two ligand AOs in each $1e$ orbital but only one ligand AO in $2a_1$. MO normalization requires smaller AO coefficients where there are more AOs. Therefore, the ligand AO coefficient in $2a_1$ will be larger, making the A-B_{ap} bond order contribution from $2a_1$ larger than individual A-B_{ba} bond order terms from the $1e$ orbitals. In other words, four ligands compete to bond through two central atom p AOs in $1e$, while one ligand bonds through one p orbital in $2a_1$. In the 12-electron AB₅ complexes $3a_1$ lengthens basal bonds and $2a_1$ shortens the apical bond. For $InCl_5^{2-}$, the ten-electron complex with square pyramidal geometry, the $3a_1$ MO is vacant but $2a_1$ still operates to make the axial bond 0.05 Å shorter than the basal bonds.

The b_1 MO of square pyramidal geometry contains no AO contribution from the apical ligand. The apical coefficient in $3a_1$ (after mixing) is much smaller than those of the basal ligands. Thus, the basal ligand AO coefficients in b_1 and $3a_1$ increase the electron density at the basal positions compared to that at the apical location. Electronegative substitutents, preferring the more electron-rich sites, occupy the basal positions in square pyramidal complexes.

The angles between apical and basal bonds are less than 90° in square pyramidal 12-electron complexes. However, in the square pyramidal ten-electron complex $InCl_5^{2-}$ the apical-basal angle is greater than 90°. Figure 12 shows roughly how the MO energies of AB₅ vary as functions of the apical-basal angle θ. The energy of $1a_1$ is nearly constant. The doubly degenerate $1e$ levels are at an energy minimum at $\theta = 90°$, where there is maximum overlap between ligand AOs and the central atom p orbitals that are perpendicular to the C_4 axis. Distortions from 90° produce a symmetric energy increase that is proportional to the cosine of the angle of distortion from 90°. For a 10° variation, cos 10° = 0.985 and the energy increase is quite small. The b_1 orbital also has a symmetric energy minimum at $\theta = 90°$. Variations from 90° push ligand AOs together out-of-phase and raise the orbital energy. The slope of the energy curve gets ever steeper the greater the angular variation from 90° and, thus, b_1 limits the amount of distortion in 10- and 12-electron complexes. The orbital $2a_1$ has an energy maximum near $\theta = 90°$ where ligand AOs are near if not on the nodal surface of the MO. For $\theta < 90°$, all ligand AOs enter $2a_1$ with the same phase; for $\theta > 90°$,

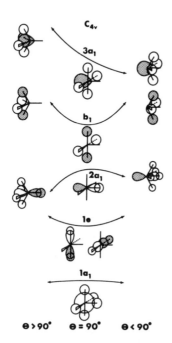

FIG. 12 How MO energies change with distortions of the apical-basal angle θ on either side of 90° for the square pyramidal AB_5 structure. [Reprinted with permission from Gimarc, *J. Am. Chem. Soc.* **100**, 2346 (1978). Copyright by the American Chemical Society.]

$\theta > 90°$ $\theta = 90°$ $\theta < 90°$

the four basal ligand AOs enter with phase opposite to that of the apical ligand AO. The energy curve of $2a_1$ is not symmetric. When all ligand AOs have the same phase, normalization requires that their coefficients be smaller than when they have different phases. Since larger coefficients produce larger energy changes, $2a_1$ has lower energy on the $\theta > 90°$ side. The angle $\theta = 90°$ is not an extremum for $3a_1$. The energy of $3a_1$ is high for $\theta > 90°$ because the basal ligands are in close out-of-phase overlap with the lone pair lobe pointing out of the base of the pyramid. As θ decreases, these out-of-phase interactions are relaxed and the energy decreases. The slope of this energy decline is steeper where the out-of-phase overlap is greater ($\theta > 90°$).

For ten-electron complexes such as $InCl_5^{2-}$ ($\theta = 108°$), the $2a_1$ MO tends to distort the square pyramid towards $\theta > 90°$. For 12-electron complexes $3a_1$ pushes the basal ligands towards $\theta < 90°$. In both cases b_1 acts to limit these distortions. The qualitative MO model provides a clear prediction of the direction of the distortion from the flat based pyramid.

RELATIVE STABILITIES OF AB_4, AB_5, AND AB_6 COMPLEXES

Figure 13 compares occupied MOs of square planar $AB_4(12e, D_{4h})$, square pyramidal $AB_5(12e, C_{4v})$, and octahedral $AB_6(14e, O_h)$ complexes. The energy is constant across $1e_u(AB_4)$-$1e(AB_5)$-$1t_{1u}(AB_6)$ because these

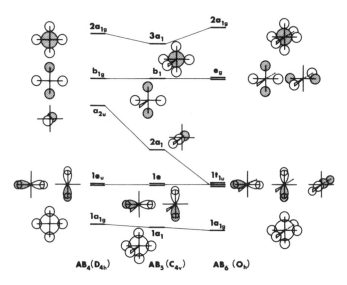

FIG. 13   Comparison of MOs of AB$_4$, AB$_5$, and AB$_6$ in square planar, square pyramidal, and octahedral shapes. [Reprinted with permission from Gimarc, *J. Am. Chem. Soc.* **100**, 2346 (1978). Copyright by the American Chemical Society.]

orbitals have identical AO compositions and coefficients. The same is true of $b_{1g}(AB_4)$-$b_1(AB_5)$-$e_g(AB_6)$. The system $1a_{1g}(AB_4)$-$1a_1(AB_5)$-$1a_{1g}(AB_6)$ is nearly constant, with the energy slightly lower at the AB$_6$ end. The energy falls much more steeply along $a_{2u}(AB_4)$-$2a_1(AB_5)$-$1t_{1u}(AB_6)$ as additional ligand AOs are added in-phase. The $3a_1(AB_5)$ orbital is stabilized by mixing with a higher energy MO of the same symmetry classification, a process not possible for the AB$_4$ and AB$_6$ $2a_{1g}$ orbitals, which consequently have higher energy than $3a_1(AB_5)$.

The highest occupied MO in 12-electron AB$_6$ complexes is the doubly degenerate nonbonding e_g pair. These MOs are composed of ligand AOs only. The energies of the e_g orbitals are determined in part by the energies of the AOs of the component ligands. For example, if the ligands are fluorines, then e_g will have lower energy than if the ligands are iodines. If the stability of the complex is related to the stability of the highest occupied MO, then there should be more fluorides than iodides among the known 12-electron AB$_6$ complexes. Of the 32 complexes listed in Table V, there are 18 fluorides, 8 chlorides, 5 bromides, and 1 iodide. If a 12-electron AB$_6$ complex is known, its fluoride is known, with chlorides, bromides, and iodides being less common in that order. The size of the central atom also determines the energy of the e_g orbitals. Although central atom AOs do not enter e_g in our model, they do determine how far apart the ligand AOs are through participation in the bonding MOs at lower energy. Larger central atoms will separate the out-of-phase overlapping ligand AOs of e_g, thereby lowering the orbital energy. This explains the increased stability of chlorides, bromides, and iodides with larger central atoms such as those at the bottom of groups IV and V. The reduction of out-of-phase overlaps among ligand AOs in e_g by smaller ligands and a larger central atom is the MO equivalent of the steric repulsion arguments of traditional valence theory that can also be used to explain these trends.

For the 14-electron AB_6 complexes, larger ligands are much more common than for the 12-electron series. Of 30 complexes in Table VI there are 8 fluorides, 8 chlorides, 7 bromides, and 7 iodides. Trends of decreasing stabilities (iodide > bromide > chloride > fluoride) have been noted for the complexes TeX_6^{2-} and BiX_6^{3-} (141,230). These trends are just the opposite of those for the 12-electron complexes and cannot be explained by simple steric repulsion arguments. The orbital $2a_{1g}(O_h)$ is occupied in the 14-electron series but empty in the 12-electron complexes. For a given central atom, larger ligands would increase the ligand-ligand AO overlap, thereby stabilizing the energy of the $2a_{1g}$ orbital. Exceptions occur if the central atom is xenon or a halogen. No chlorides, bromides, or iodides of xenon or halogen central atoms have been prepared in the 14-electron AB_6 series. Ligands must be more electronegative than the central atom to which they are bound. This must be so because MO nodes run through or near the central atom, pushing electron density away from the central atom and towards the ligands. Recall that the energy of the e_g orbitals is largely determined by the ligand AO energies. Placing the more electronegative element at the center and using the less electronegative element as ligands would increase the e_g energy (231). For xenon and halogen central atoms, only fluorine is electronegative enough to serve as a ligand.

In the 12-electron square pyramidal AB_5 complexes, the highest occupied MO is $3a_1(C_{4v})$. Although this orbital is A–B antibonding, the in-phase overlaps among ligand AOs tend to stabilize large ligand complexes. The similar nature of the highest occupied MOs in $AB_5(12e)$ and $AB_6(14e)$ suggests that these two series of complexes might be formed from the same elements and exhibit the same relative stabilities. A comparison of Tables VI and IX shows that this is roughly true. There are 30 known $AB_6(14e, O_h)$ complexes and 20 $AB_5(12e, C_{4v})$ complexes. Between these two series there are 18 related pairs composed of the same elements, such as $TeBr_5^-$ and $TeBr_6^{2-}$. In other words, for all but two of the known $AB_5(12e, C_{4v})$ complexes, the related $AB_6(14e, O_h)$ complex is also known. The similar electronic structures favor complexes of the same elements.

The highest occupied MO of the 12-electron square planar AB_4 complexes is $2a_{1g}$, which is also A–B antibonding with in-phase ligand-ligand interactions. Table III contains only 6 $AB_4(12e, D_{4h})$ complexes of which there are 4 fluorides and 2 chlorides. The distribution of substituents in the AB_4 series is not comparable to that of the seemingly related AB_5 and AB_6 complexes. Notice that the number of in-phase ligand-ligand overlapping pairs in the highest occupied MO decreases from AB_6 (12 pairs) to AB_5 (8 pairs) to AB_4 (4 pairs). Below the antibonding MO lie nonbonding orbitals that have ligand-ligand out-of-phase overlapping pairs. For $AB_4(12e, D_{4h})$ the 4 pairs of ligand-ligand out-of-phase overlaps in b_{1g} cancel the in-phase interactions in $2a_{1g}$, nearly eliminating this entire series of complexes and causing these complexes to be composed of elements different from those of the $AB_5(12e, C_{4v})$ and $AB_6(14e, O_h)$ series.

The ten-electron $AB_4(C_{2v})$ complexes fill out nearly the same portions of the periodic table as do the $AB_5(12e, C_{4v})$ and $AB_6(14e, O_h)$ series. Compare Tables II, VI, and IX. Among the 32 $AB_4(10e, C_{2v})$ and 30 $AB_6(14e, O_h)$ complexes there are 23 related pairs such as $SeBr_4$ and $SeBr_6^{2-}$. Comparing the 32 $AB_4(10e, C_{2v})$ and 20 $AB_5(12e, C_{4v})$ complexes, there are 19 related pairs such as SbI_4^- and SbI_5^{2-}. The AB_4 series includes 9 fluorides, 9 chlorides, 8 bromides, and 6 iodides, a sizable proportion of large ligand complexes. EMF and solubility studies of the tetrahalide complexes BiX_4^- and TlX_4^{3-} have been interpreted as demonstrating the decreasing strength of these complexes in the order $I^- > Br^- > Cl^-$ (35,230).

Figure 2 shows that the highest occupied MO in $AB_4(10e, C_{2v})$ complexes is $3a_1$, an A-B antibonding MO related to $3a_1(AB_5, C_{4v})$ and $2a_{1g}(AB_4, D_{4h}$ and $AB_6, O_h)$. The 5 pairs of in-phase overlaps among ligands in $3a_1(AB_4, C_{2v})$ provide a rotationalization of the stability of large ligand complexes in this AB_4 series. The orbital related to the nonbonding MOs $b_{1g}(AB_4, D_{4h})$, $b_1(AB_5, C_{4v})$, and $e_g(AB_6, O_h)$ is empty in the diagonally folded square AB_4 complexes. Figure 2 shows how the A-B nonbonding MO $b_{1g}(AB_4, D_{4h})$ forms below the highest occupied orbital $2a_{1g}(D_{4h})$ as the diagonally folded square (C_{2v}) structure of AB_4 is flattened to square planar shape.

Figure 14 compares the occupied MOs of $AB_4(C_{2v})$, $AB_5(D_{3h})$, and $AB_6(O_h)$. There is much less similarity between MOs of these different structures than we found in Fig. 13. Again, the nodeless MO system $1a_1(C_{2v})$-$1a_1'(D_{3h})$-$1a_{1g}(O_h)$ is nearly constant in energy. The MO $1b_1$-$1a_2''$-$1t_{1u}$ is composed of orbitals of identical energy and AO composition. Assuming an angle of 120° between the equatorial ligands of the diagonally folded square (C_{2v}) structure gives $1b_2(C_{2v})$ and one of the components of $1e'(D_{3h})$ identical AO compositions, overlaps, and energies. Among the antibonding $3a_1(C_{2v})$-$3a_1'(D_{3h})$-$2a_{1g}(O_h)$ system, the $3a_1(C_{2v})$ MO has the lowest energy because it can mix with the higher energy $4a_1$ MO, while $3a_1'$ and $2a_{1g}$ are the highest energy MOs of their symmetries in their respective systems.

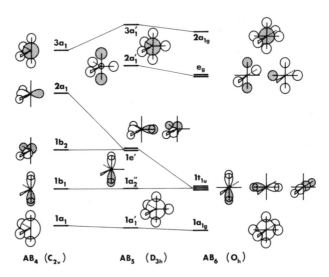

FIG. 14   Comparison of MOs of AB_4, AB_5, and AB_6 in folded square, trigonal bipyramidal, and octahedral shapes. [Reprinted with permission from Gimarc, *J. Am. Chem. Soc.* **100**, 2346 (1978). Copyright by the American Chemical Society.]

The $AB_4(10e, C_{2v})$ complexes are often reviewed along with those of $AB_5(10e, D_{3h})$ series, the structure of the AB_4 complexes being described as trigonal bipyramidal like AB_5 but with a lone pair of electrons occupying one of the equatorial positions. Despite their geometrical similarity, the electronic structures of the two classes of complexes are different enough to make the kinds of complexes contained in the two classes quite different. Consider the differences in the highest occupied MOs at the ten-electron level. In $3a_1(AB_4)$ the axial and equatorial ligand AOs have the same phase that tends to stabilize complexes with large ligands. On the other hand, $2a_1'(AB_5)$ has axial and equatorial ligand AOs of opposite phase, an arrangement stabilized by small ligands and large central atoms, a conclusion supported by the pattern of known ten-electron AB_5 complexes shown in Table VIII. Of the 18 AB_5 complexes in Table VIII, there are 9 fluorides, 7 chlorides, 2 bromides, and no iodides. The data here are rather sparse. Only 18 AB_5 complexes with ten valence electrons are known compared to 32 AB_4 complexes with ten electrons. Between these two classes of complexes there are only 6 related pairs such as PF_4^-, PF_5.

There are no well-characterized ten-electron AB_5 complexes in which the central atom A is an element from the groups to the right of V in the periodic table. SF_5^+ is an example of a complex missing from Table VIII. Now SF_5^+ has been produced in the gas phase both by photon or electron ionization of SF_6 (*232*) or by the reaction of SF_6 with ions of high electron affinity such as He^+ (*233*). In both cases it is presumed that the SF_6^+ ion formed decomposes spontaneously into SF_5^+ and F. Comparison of the MOs for $AB_5(10e, D_{3h})$ and $AB_6(12e, O_h)$ in Fig. 14 suggests why SF_5^+ might be unstable relative to SF_6 in solution. The $1e'$ and $2a_1'$ MOs of $AB_5(D_{3h})$ have slightly higher energy than the related $1t_{1u}$ and e_g MOs of $AB_6(O_h)$. For the process $AB_5(10e, D_{3h}) + :B \rightarrow AB_6(12e, O_h)$, the electron pair from the incoming ligand goes into one of the nonbonding e_g orbitals of the product. The nonbonding e_g MO of the complex is higher in energy than the nonbonding AO of the incoming ligand, but at least the electron pair does not have to go into an antibonding MO of the complex. If the central atom of the AB_5 complex is from a group to the right of V, then the ten-electron complex will be positively charged and electrostatics will favor the formation of the AB_6 complex.

RELATIVE STABILITIES OF AB_4 COMPLEXES

Table I lists 45 known eight-electron AB_4 halide complexes of the representative elements considered here. Although the complexes of elements in groups III and IV are complete, in the sense that fluorides, chlorides, bromides, and iodides have all been prepared, only a few eight-electron complexes are known for the group V elements. For example, NF_4^+ has been prepared but not NCl_4^+, NBr_4^+, nor NI_4^+. Conventional valence theory explains this by steric repulsion arguments. Large ligands such as Cl, Br, or I would be too close together if attached to a central atom as small as N. Translating this into MO terms, only bonding MOs are occupied in the eight-electron tetrahedral AB_4 complexes but the $1t_1$ bonding MOs contain ligand-ligand antibonding interactions as well. Consider the member **g** of the $1t_1(T_d)$ set (**14**). Each ligand AO in **g** has two out-of-phase overlaps

g

14

with other ligand AOs. If ligand AOs are of small radius and the central atom is large, all ligand-ligand overlaps are small and **g** is net bonding because of ligand AO overlaps with the adjacent lobes of the central atom p orbital. However, for large ligands and small central atoms the out-of-phase ligand-ligand interactions take over to produce a net nonbonding or antibonding MO. Although NCl_4^+ is not known, chlorides of the larger atoms P, As, and Sb have been prepared. Therefore, it is surprising that AsF_4^+ and SbF_4^+ are missing from Table I.

Why are the atoms of the second and lower rows of the periodic table hypervalent (234)? In other words, why do we have SF_6 and SeF_6 but not OF_6? Conventional valence theory usually gives an answer in terms of central atom size. There is not enough room around an oxygen atom to accommodate six ligands even though they might be as small as fluorine atoms. Sulfur and selenium are larger and, therefore, a larger number of ligands can fit around them. Another explanation involves the participation in bonding of empty d AOs that are close to the occupied valence orbitals in S and Se but too high in energy on O. The following explanation also involves central atom size, but it is based on MO arguments and does not require participation of d orbitals.

The stability of three-, four-, five-, and six-coordinate complexes with more than eight valence electrons requires that two or more electrons occupy nonbonding or even antibonding MOs. These MOs usually contain out-of-phase interactions between component ligand AOs. In OF_6 and SF_6, for example, these MOs would be the e_g pair **a** and **b**. Large central atoms tend to keep these MOs nonbonding; small central atoms allow the out-of-phase ligand AOs to approach each other and raise the MO energy.

The representations of the $1a_2''$ and $1e'$ (AB_6, D_{3h}) orbitals in Fig. 7 show that ligand-ligand AOs overlap in-phase with adjacent lobes of the central atom p AO to make these the bonding MOs they are normally considered to be. However, those figures show that these orbitals also have ligand-ligand out-of-phase interactions. If the central atom were small and the ligands large, then these ligand-ligand out-of-phase interactions could destabilize the $1a_2''$ and $1e'$ MOs. This apparently happens to eliminate the class of eight-electron trigonal prismatic AB_6 complexes.

RELATIVE A-B BOND DISTANCES

There is a remarkable similarity among the occupied MOs of 14-electron octahedral AB_6, 12-electron square planar AB_4, and 10-electron linear AB_2 complexes. Because of this orbital relationship, one might expect the A-B bond distances to be equal through a series of complexes composed of the same kinds of atoms. Excellent experimental data are available for the xenon fluorides. The Xe-F distances as measured by different techniques are 1.890 ± 0.005 Å in XeF_6 (gas phase electron diffraction), 1.951 and 1.954 ± 0.002 Å in XeF_4 (neutron diffraction in crystal), and 1.977 ± 0.002 Å in XeF_2 (gas phase high resolution infrared spectrum) (235). Although the complexes are not in the same physical state and the experimental techniques are all different, the differences in bond distances are more than ten times the stated uncertainties in the results. The trend in bond distances can be accounted for as a result of differences in A-B antibonding interactions between the highest occupied MOs of these complexes.

Individual MO bond orders are proportional to A-B AO overlaps multiplied by the product of the coefficients of the overlapping AOs. Except for the bonding MOs $1a_{1g}(AB_6)$, $1a_{1g}(AB_4)$, and $1\sigma_g(AB_2)$ and the antibonding MOs $2a_{1g}(AB_6)$, $2a_{1g}(AB_4)$, and $2\sigma_g(AB_2)$, all other sets of occupied MOs are either nonbinding, and hence contribute nothing to the calculated bond orders, or have identical AO compositions and coefficients, which require them to make identical contributions to calculated A-B bond orders. MO normalization requires each MO of the bonding set $1a_{1g}(AB_6)$, $1a_{1g}(AB_4)$, and $1\sigma_g(AB_2)$ to have smaller AO coefficients than its corresponding member of the antibonding set $2a_{1g}$, $2a_{1g}$, and $2\sigma_g$, which therefore must determine bond order differences by virtue of their larger AO coefficients and larger contributions to the total calculated bond order (15). Because

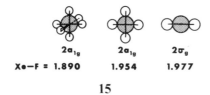

$2a_{1g}$	$2a_{1g}$	$2\sigma_g$
Xe–F = 1.890	1.954	1.977

15

there are fewer ligands in AB_4 than in AB_6, MO normalization requires that ligand AO coefficients be larger in $2a_{1g}(AB_4)$ than in $2a_{1g}(AB_6)$. For the same reason, the ligand coefficients in $2\sigma_g(AB_2)$ should be still larger. Therefore, individual bond order contributions should increase in magnitude from XeF_6 to XeF_4 to XeF_2. Since the contributions are antibonding in each of these MOs, the increase in magnitude weakens the A-B bond so that bond lengths should increase from XeF_6 to XeF_4 to XeF_2 as observed. Similar trends should occur in the following series of known complexes for which bond distances in one or more members have not been measured. Triplets: ClF_6^-, ClF_4^-, ClF_2^-; BrF_6^-, BrF_4^-, BrF_2^-; pairs: IF_6^-, IF_4^-; $PoCl_6^{2-}$, $PoCl_4^{2-}$.

There is a comparable similarity between the occupied MOs of 12-electron square pyramidal AB_5 and 10-electron T-shaped AB_3 complexes. Because of the larger number of phase differences in the antibonding highest occupied MOs $3a_1(AB_5)$ and $3a_1(AB_3)$, these orbitals have larger AO coefficients than lower energy MOs and they therefore determine the differences in A-B bond orders (16). MO normalization makes the ligand AO coefficients larger for the less substituted complex. Since the A-B interactions are antibonding in both $3a_1$ orbitals, the larger coefficients in $3a_1(AB_3)$ produce longer A-B bonds in this complex. The experimental data in **16** for XeF_5^+ and XeF_3^+ (236) show the longer bonds in the less substituted

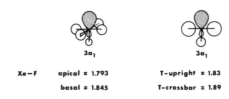

16

complex. Similar comparisons for BrF$_5$ and BrF$_3$ (237) confirm the trend. The same trend is to be expected in the following known pairs for which bond distances of one or both members have not been reported: ClF$_5$, ClF$_3$; SeCl$_5^-$, SeCl$_3^-$; SeBr$_5^-$, SeBr$_3^-$.

IONIC EQUILIBRIA

SeBr$_4$, SeCl$_4$, and TeX$_4$ (X = F, Cl, Br, and I) are known to have ionic or slightly ionic structures AB$_3^+ \cdots$ B$^-$ in the solid state (47,48,49a,238) and in some cases in solution (49b). The planar (D_{3h}) group III trihalides can add another halide ion to form tetrahedral complexes. Trigonal pyramidal (C_{3v}) group V trihalides add one more halide ligand to yield the folded square tetrahalides. The square planar tetrafluorides are synthesized by adding fluoride ion to the T-shaped (C_{2v}) group VII trifluorides. The tetrahalide complexes of group IV and V add halide ions to produce trigonal bipyramidal (D_{3h}) complexes (239). The diagonally folded square tetrahalides of elements in groups VI and VII add halide ligands to form square pyramidal (C_{4v}) pentahalides (59–61). Trigonal bipyramidal group V pentahalides add halide ligands to form octahedral complexes (239). The ions XF$_6^-$ (X = Cl, Br, or I) are formed by the reaction of XF$_5$ and F$^-$ (145–147). In the solid, XeF$_6$ apparently dissociates into the ions XeF$_5^+$ and F$^-$ (240). A similar dissociation is believed to occur in HF solution (241). Lewis acids and XeF$_6$ are known to form highly stable salts of XeF$_5^+$ (242). Raman spectra of SeBr$_6^{2-}$ solutions indicate an equilibrium between SeBr$_6^{2-}$ and SeBr$_5^-$ + Br$^-$ (140). Equations [1]–[7] summarize these processes. The equations include the number of valence electrons and the molecular geometry of each complex. The symbol over the arrow indicates the assumed transition state symmetry.

$$AB_3^+(6e, D_{3h}) + B^-(2e) \xrightarrow{C_{3v}} AB_4(8e, T_d), \qquad [1]$$

$$AB_3^+(8e, C_{3v}) + B^-(2e) \xrightarrow{C_s} AB_4(10e, C_{2v}), \qquad [2]$$

$$AB_3(10e, C_{2v}) + B^-(2e) \xrightarrow{C_{2v}} AB_4^-(12e, D_{4h}), \qquad [3]$$

$$AB_4(8e, T_d) + B^-(2e) \xrightarrow{C_{2v}} AB_5^-(10e, D_{3h}), \qquad [4]$$

$$AB_4(10e, C_{2v}) + B^-(2e) \xrightarrow{C_{2v}} AB_5^-(12e, C_{4v}), \qquad [5]$$

$$AB_5(10e, D_{3h}) + B^-(2e) \xrightarrow{C_{2v}} AB_6^-(12e, O_h), \qquad [6]$$

$$AB_5(12e, C_{4v}) + B^-(2e) \xrightarrow{C_{4v}} AB_6^-(14e, O_h). \qquad [7]$$

Not surprisingly, MO correlation diagrams show that all these processes are symmetry allowed. The diagrams are easy to construct and two of them are included here as examples.

Figure 15 correlates reactant and product MOs for Eq. [5]. Energies decrease along $1b_2(AB_4)$-$1e(AB_5)$ and $2a_1(AB_4)$-$2a_1(AB_5)$ and increase from $3a_1(AB_4)$ to $3a_1(AB_5)$. The electron pair of the reacting ion B^- must rise in energy as it goes into the nonbonding b_1 MO of AB_5^-. Qualitative MO arguments alone cannot predict which side is favored. Electrostatic repulsions oppose the addition of a negatively charged ligand to an already negatively charged complex reactant. There are 13 cases in which an $AB_4(10e, C_{2v})$ complex is known, but the related AB_5^- ($12e, C_{4v}$) complex is not. Of these 13 cases, 9 would require the addition of a negatively charged

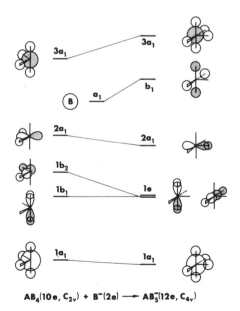

$$AB_4(10e, C_{2v}) + B^-(2e) \longrightarrow AB_5^-(12e, C_{4v})$$

Fig. 15 Orbital correlations for Eq. [5]: $AB_4(10e, C_{2v}) + B^-(2e) \rightarrow AB_5^-(12e, C_{4v})$. [Reprinted with permission from Gimarc, *J. Am. Chem. Soc.* **100**, 2346 (1978). Copyright by the American Chemical Society.]

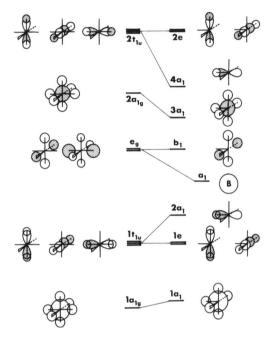

FIG. 16   Orbital correlations for Eq. [7]: $AB_5(12e, C_{4v}) + B^-(2e) \to AB_6^-(14e, O_h)$.

ligand to an already negatively charged AB_4 complex. For example, PF_4^- exists but PF_5^{2-} is unknown. Such additions do occur however when the complex is large enough to diffuse the extra charge. Examples are $PbBr_4^{2-}$, $PbBr_5^{3-}$ and $BiCl_4^-$, $BiCl_5^{2-}$.

Figure 16 is a correlation diagram for the process in Eq. [7]. The energy of the electron pair of the incoming ligand B^- rises as it fills the antibonding e_g MO of the AB_6^- product. The energy of $2a_{1g}(AB_6)$ is higher than that of $3a_1(AB_5)$ because the $3a_1(AB_5)$ orbital is stabilized by mixing with a higher MO of the same symmetry. For most related pairs, electrostatic repulsions oppose further ligand additions such as $TeF_5^- + F^- \to TeF_6^{2-}$. The fact that so many related AB_5, AB_6^- pairs are known must mean that equilibrium between the two complex forms is fairly balanced. The diagram for Eq. [6] is very similar to Fig. 16. In this case, the AB_5 reactant has trigonal bipyramidal shape but the pattern of energy levels is the same.

MOLECULAR HALOGEN REACTIONS

As solid $SeCl_4$ vaporizes it dissociates completely into gaseous $SeCl_2$ and Cl_2 (243). Solid $SeBr_4$ similarly dissociates on vaporization (244) and on dissolving in a wide range of solvents (245). The hexafluorides of sulfur, selenium, and tellurium are ordinarily prepared by burning these elements in fluorine (246). By carefully controlling the conditions (0° or less, dilute fluorine) these reactions can be made to produce only the tetrafluorides (247). SF_5Cl can be prepared by the direct reaction of SF_4 and ClF at 350°C and 5 atm (248). The same product can be formed much more readily and under milder conditions if SF_4 and CsF are first combined to form the adduct $Cs^+SF_5^-$, which is then reacted with ClF forming SF_5Cl and re-generating CsF (249). It is clear that the addition of the third halogen di-atomic to the tetrafluorides is more difficult than the addition of the first two. Consider the processes:

$$AB_4(10e, C_{2v}) \xrightarrow{\ C_{2v}\ } AB_2(8e, C_{2v}) + B_2(2e), \qquad [8]$$

$$AB_6(12e, O_h) \longrightarrow AB_4(10e, C_{2v}) + B_2(2e). \qquad [9]$$

The dissociation in Eq. [8] is symmetry allowed as shown in Fig. 17. The bonding σ_g MO of product B_2 has been placed below the bonding AB_2 levels that contain nodes. It could be even higher than $2a_1(AB_2, C_{2v})$ without changing the allowed nature of the process. As equatorial ligands of AB_4 depart, the bonding b_2 MO of AB_4 becomes nonbonding b_2 in AB_2. Since the eight-electron AB_2 product is bent, the energy of $1b_1(AB_2)$ is higher than that of $1b_1(AB_4)$. The antibonding $3a_1$ orbital of the AB_2 product is empty.

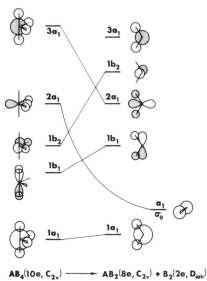

$AB_4(10e, C_{2v}) \longrightarrow AB_2(8e, C_{2v}) + B_2(2e, D_{\infty h})$

FIG. 17   Orbital correlations for the dissociation $AB_4(C_{2v}) \to AB_2(C_{2v}) + B_2$.

Figure 18 shows that the dissociation of Eq. [9] is symmetry forbidden. For simplicity, in constructing Fig. 18 we assume the angle between the two equatorial ligands of AB_4 is fixed at 90°. The bonding MO of the product B_2 (a_1 or σ_g) is assumed to have an energy comparable to that of the nodeless $1a_1$ MO of AB_4 and below that of the group of AB_4 orbitals that are related in AO composition to the $1t_{1u}$ MOs of AB_6. The antibonding MO (b_2 or σ_u) of the B_2 product has been placed below the antibonding $2b_2$ MO of the AB_4 product, but the order of these two orbitals might conceivably be reversed depending on the particular system considered. With either ordering of the antibonding b_2 levels, the dissociation of 12-electron AB_6 complexes is symmetry forbidden because electrons would flow from the e_g orbital of reactant into either the σ_u (b_2) MO of B_2 as shown or the $2b_2$ MO of AB_4, leaving $3a_1(AB_4)$ vacant at lower energy. If the order of antibonding b_2 levels of products is that shown in Fig. 17, then the dissociation of 14-electron AB_6 complexes is also symmetry forbidden. Empirical evidence suggests that this is the case. The 14-electron complexes are synthesized by the addition of F^- to AB_5 rather than by direct addition of a diatomic halogen molecule to AB_4.

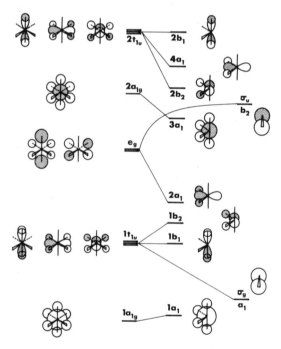

FIG. 18 Orbital correlations for the dissociation $AB_6(O_h) \rightarrow AB_4(C_{2v}) + B_2$. [Reprinted with permission from Gimarc et al., J. Am. Chem. Soc. **100**, 2334 (1978). Copyright by the American Chemical Society.]

REFERENCES

AB_4, Eight-Electron Complexes

(1) BF_4^- : G. Burton, *Acta Crystallogr.* **B24**, 1703 (1968).
(2) BCl_4^- : O. Glemser, B. Krebs, J. Wegener, and E. Kindler, *Angew. Chem.* **81**, 568 (1969).
(3) BBr_4^- : J. A. Creighton, *J. Chem. Soc.* 6589 (1965).
(4) BI_4^- : I. Y. Ahmed and C. D. Schmulbach, *Inorg. Chem.* **8**, 1411 (1969).
(5) AlF_4^- : N. A. Matwiyoff and W. E. Wageman, *Inorg. Chem.* **9**, 1031 (1970).
(6) $AlCl_4^-$: N. C. Baenziger, *Acta Crystallogr.* **4**, 216 (1951).
(7) $AlBr_4^-$, AlI_4^- : G. M. Begun, C. R. Boston, G. Torsi, and G. Mamantov, *Inorg. Chem.* **10**, 886 (1971); H. Haraguchi and S. Fujiwara, *J. Phys. Chem.* **73**, 3467 (1969).
(8) GaF_4^- : L. K. Beck, B. H. Kugler, and H. M. Haendler, *J. Solid State Chem.* **8**, 312 (1973); P. Bukovec, B. Orel, and J. Siftar, *Monatsh. Chem.* **104**, 194 (1973).
(9) $GaCl_4^-$, $GaBr_4^-$, GaI_4^- : A. Fratiello, R. E. Lee, and R. E. Schuster, *Inorg. Chem.* **9**, 82 (1970).
(10) InF_4^- : T. Ryhl, *Acta Chem. Scand.* **23**, 2667 (1969).
(11) $InCl_4^-$, $InBr_4^-$, InI_4^- : D. H. Brown and D. T. Stewart, *J. Inorg. Nucl. Chem.* **32**, 3751 (1970).
(12) $TlCl_4^-$, $TlBr_4^-$, TlI_4^- : K. Schmidt, *J. Inorg. Nucl. Chem.* **32**, 3549 (1970).
(13) CF_4 : C. W. W. Hoffman and R. L. Livingston, *J. Chem. Phys.* **21**, 565 (1953).
(14) CCl_4 : Y. Morino, Y. Nakamura, and T. Iijima, *J. Chem. Phys.* **32**, 643 (1960).
(15) CBr_4, CI_4 : M. W. Lister and L. E. Sutton, *Trans. Faraday Soc.* **37**, 393 (1941).
(16) SiF_4 : B. Beagley, D. P. Brown, and J. M. Freeman, *J. Mol. Struct.* **18**, 337 (1973); K. Hagen and K. Hedberg, *J. Chem. Phys.* **59**, 1549 (1973).
(17) $SiCl_4$: Y. Morino and Y. Murata, *Bull. Chem. Soc. Jpn.* **38**, 104 (1965).
(18) $SiBr_4$, SiI_4 : M. W. Lister and L. E. Sutton, *Trans. Faraday Soc.* **37**, 393 (1941).
(19) GeF_4 : A. D. Gaunt, M. Mackle, and L. E. Sutton, *Trans. Faraday Soc.* **47**, 943 (1951).
(20) $GeCl_4$: Y. Morino, Y. Nakamura, and T. Iijima, *J. Chem. Phys.* **32**, 643 (1960).
(21) $GeBr_4$: G. B. B. Souza and J. D. Wieser, *J. Mol. Struct.* **25**, 442 (1975).
(22) GeI_4 : M. W. Lister and L. E. Sutton, *Trans Faraday Soc.* **37**, 393 (1941).
(23) SnF_4 : R. Hoppe and W. Dähne, *Naturwissenschaften* **49**, 254 (1962).
(24) $SnCl_4$: H. Fujii and M. Kimura, *Bull. Chem. Soc. Jpn.* **43**, 1933 (1970).
(25) $SnBr_4$, SnI_4 : M. W. Lister and L. E. Sutton, *Trans. Faraday Soc.* **37**, 393 (1941).
(26) PbF_4 : G. Nagarajan, *Bull. Soc. Chim. Belges* **71**, 119 (1962).
(27) $PbCl_4$: M. W. Lister and L. E. Sutton, *Trans. Faraday Soc.* **37**, 393 (1941).
(28) $PbBr_4$, PbI_4 : G. Nagarajan, *Bull Soc. Chim. Belges* **71**, 119 (1962).
(29) NF_4^+ : K. O. Christe, J. P. Guertin, and A. E. Pavlath, *Inorg. Nucl. Chem. Lett.* **2**, 83 (1966); W. E. Tolberg, R. T. Rewick, R. S. Stringham, and M. E. Hill, *Inorg. Nucl. Chem. Lett.* **2**, 79 (1966).
(30) PF_4^+ : G. S. H. Chen and J. Passmore, *Chem. Commun.* 559 (1973).
(31) PCl_4^+ : P. H. Collins and M. Webster, *Acta Crystallogr.* **B28**, 1260 (1972).
(32) PBr_4^+ : G. L. Breneman and R. D. Willett, *Acta Crystallogr.* **23**, 467 (1967).
(33) $AsCl_4^+$: H. Preiss, *Z. Anorg. Allg. Chem.* **380**, 71 (1971).
(34) $SbCl_4^+$: H. B. Miller, H. W. Baird, C. L. Bramblett, and W. K. Templeton, *Chem. Commun.* 262 (1972); H. Preiss, *Z. Anorg. Allg. Chem.* **389**, 254 (1972).

AB_4, Ten-Electron Complexes

(35) TlI_4^{3-} : R. O. Nilsson, *Arkiv Kemi* **10**, 363 (1957).
(36) $SnCl_4^{2-}$: J. Itoh, R. Kusaka, Y. Yamagata, R. Kiriyama, H. Ibamoto, T. Kanda, and Y. Masuda, *J. Phys. Soc. Jpn.* **8**, 287, 293 (1953).

(37) $PbCl_4^{2-}$, $PbBr_4^{2-}$, PbI_4^{2-} : A. J. Macfarlane and R. J. P. Williams, *J. Chem. Soc. A* 1517 (1969).

(38) PF_4^- : T. C. Rhyne and J. G. Dillard, *Inorg. Chem.* **10**, 730 (1971).

(39) PBr_4^- : K. B. Dillon and T. C. Waddington, *Chem. Commun.* 1317 (1969).

(40) AsF_4^- : T. C. Rhyne and J. G. Dillard, *Inorg. Chem.* **13**, 322 (1974).

(41) $AsCl_4^-$, $AsBr_4^-$: G. Y. Ahlijah and M. Goldstein, *J. Chem. Soc. A* 326 (1970).

(42) SbF_4^- : C. J. Adams and A. J. Downs, *J. Chem. Soc. A* 1534 (1971).

(43) $SbCl_4^-$, $SbBr_4^-$, SbI_4^-, $BiCl_4^-$, $BiBr_4^-$, BiI_4^- : G. Y. Ahlijah and M. Goldstein, *J. Chem. Soc. A* 326 (1970).

(44) SF_4 : W. M. Tolles and W. D. Gwinn, *J. Chem. Phys.* **36**, 1119 (1962).

(45) SCl_4 : N. Wiberg, G. Schwenk, and K. H. Schmid, *Chem. Ber.* **105**, 1209 (1972).

(46) SeF_4 : I. C. Bowater, R. D. Brown, and F. R. Burden, *J. Mol. Spectrosc.* **28**, 454 (1968).

(47) $SeCl_4$, $SeBr_4$: G. C. Hayward and P. J. Hendra, *J. Chem. Soc. A* 643 (1967); N. Katsaros and J. W. George, *Inorg. Chim. Acta* **3**, 165 (1969).

(48) TeF_4 : A. J. Edwards and F. I. Hewaidy, *J. Chem. Soc. A* 2977 (1968).

(49) $TeCl_4$: (a) B. Buss and B. Krebs, *Inorg. Chem.* **10**, 2795 (1971); (b) D. M. Adams and P. J. Lock, *J. Chem. Soc. A* 145 (1967); (c) N. Katsaros and J. W. George, *Inorg. Chim. Acta* **3**, 165 (1969).

(50) $TeBr_4$: D. M. Adams and P. J. Lock, *J. Chem. Soc. A* 145 (1967); N. Katsaros and J. W. George, *Inorg. Chim. Acta* **3**, 165 (1969).

(51) TeI_4 : B. Krebs and V. Paulat, *Acta Crystallogr.* **B32**, 1470 (1976).

(52) $PoCl_4$: K. W. Bagnall, R. W. M. D'Eye, and J. H. Freeman, *J. Chem. Soc.* 2320 (1955).

(53) $PoBr_4$: K. W. Bagnall, R. W. M. D'Eye, and J. H. Freeman, *J. Chem. Soc.* 3959 (1955).

(54) PoI_4 : K. W. Bagnall, R. W. M. D'Eye, and J. H. Freeman, *J. Chem. Soc.* 3385 (1956).

(55) ClF_4^+ : K. O. Christe and W. Sawodny, *Inorg. Chem.* **12**, 2879 (1973); K. O. Christe, J. F. Hon, and D. Pilipovich, *Inorg. Chem.* **12**, 84 (1973).

(56) BrF_4^+ : M. D. Lind and K. O. Christe, *Inorg. Chem.* **11**, 608 (1972).

(57) IF_4^+ : H. W. Baird and H. F. Giles, *Acta Crystallogr.* **25A**, S115 (1969); D. D. Gibler, *Diss. Abst. Int. B* **34**, 3688 (1974).

AB_4, *12-Electron Complexes*

(58) $PoCl_4^{2-}$: K. W. Bagnall and J. H. Freeman, *J. Chem. Soc.* 2770 (1956).

(59) ClF_4^- : K. O. Christe and J. P. Guertin, *Inorg. Chem.* **5**, 473 (1966); K. O. Christe and W. Sawodny, *Z. Anorg. Allg. Chem.* **357**, 125 (1968).

(60) BrF_4^- : A. J. Edwards and G. R. Jones, *J. Chem. Soc. A* 1936 (1969); K. O. Christe and C. J. Schack, *Inorg. Chem.* **9**, 1852 (1970).

(61) IF_4^- : K. O. Christe and D. Naumann, *Inorg. Chem.* **12**, 59 (1973).

(62) ICl_4^- : R. C. L. Mooney, *Z. Kristallogr.* **98**, 377 (1938); R. J. Elema, J. L. de Boer, and A. Vos, *Acta Crystallogr.* **16**, 243 (1963).

(63) XeF_4 : J. H. Burns, P. A. Agron, and H. A. Levy, *Science* **139**, 1208 (1963).

AB_4, *Oxides and Sulfides*

(64) XeO_4 : G. Gundersen, K. Hedberg, and J. L. Huston, *J. Chem. Phys.* **52**, 812 (1970).

(65) SiS_4^{4-} : J. T. Lemley, *Acta Crystallogr.* **B30**, 549 (1974).

(66) SnO_4^{4-} : R. Marchand, Y. Piffard, and M. Tournoux, *Acta Crystallogr.* **B31**, 511 (1975).

(67) SnS_4^{4-} : S. Jaulmes, J. Rivet, and P. Laruelle, *Acta Crystallogr.* **B33**, 540 (1977).

(68) PS_4^{3-}, AsS_4^{3-}, SbS_4^{3-} : A. Müller, E. Diemann, and M. J. F. Leroy, *Z. Anorg. Allg. Chem.* **372**, 113 (1970); M. Palazzi, S. Jaulmes, and P. Laruelle, *Acta Crystallogr.* **B30**, 2378 (1974).

AB_4, *Nine-Electron Radicals*

(69) PO_4^{4-}: M. C. R. Symons, *J. Chem. Phys.* **53**, 857 (1970).
(70) AsO_4^{4-}: M. Hampton, F. G. Herring, W. C. Lin, and C. A. McDowell, *Mol. Phys.* **10**, 565 (1966).
(71) SeO_4^{3-}: K. V. S. Rao and M. C. R. Symons, *J. Chem. Soc. Dalton* 147 (1972).
(72) ClO_4^{2-}: M. B. D. Bloom, R. S. Eachus, and M. C. R. Symons, *J. Chem. Soc. A* 1235 (1970).
(73) CBr_4^{-}: S. P. Mishra and M. C. R. Symons, *Chem. Commun.* 577 (1973).
(74) SiF_4^{-}: J. R. Morton and K. F. Preston, *Mol. Phys.* **30**, 1213 (1975).
(75) PF_4: (a) R. W. Fessenden and R. H. Schuler, *J. Chem. Phys.* **45**, 1845 (1966); (b) W. Nelson, G. Jackel, and W. Gordy, *J. Chem. Phys.* **52**, 4572 (1970).
(76) PCl_4: G. F. Kokoszka and F. E. Brinckman, *J. Am. Chem. Soc.* **92**, 1199 (1970); S. P. Mishra and M.C. R. Symons, *J. Chem. Soc. Dalton* 139 (1976).
(77) SF_4^{+}: R. W. Fessenden and R. H. Schuler, *J. Chem. Phys.* **45**, 1845 (1966).
(78) $POCl_3^{-}$: A. Begum and M. C. R. Symons, *J. Chem. Soc. A* 2065 (1971).
(79) $SO_2Cl_2^{-}$: T. Gillbro and F. Williams, *Chem. Phys. Lett.* **20**, 436 (1973); K. V. S. Rao and M. C. R. Symons, *J. Chem. Soc. Dalton* 9 (1973).
(80) E. L. Muetterties and R. A. Schunn, *Quart. Rev.* **20**, 245 (1966).

AB_4, *Ten-Electron Mixed-Ligand Complexes*

(81) $SeO_2F_2^{2-}$: R. J. Gillespie, P. Spekkens, J. B. Milne, and D. Moffett, *J. Fluorine Chem.* **7**, 43 (1976).
(82) $TeO_2F_2^{2-}$: J. B. Milne and D. Moffett, *Inorg. Chem.* **12**, 2240 (1973).
(83) $ClO_2F_2^{-}$: K. O. Christe and E. C. Curtis, *Inorg. Chem.* **11**, 35 (1972).
(84) $BrO_2F_2^{-}$: R. J. Gillespie and P. Spekkens, *J. Chem. Soc. Dalton* 2391 (1976).
(85) $IO_2F_2^{-}$: L. Helmholz and M. T. Rogers, *J. Am. Chem. Soc.* **62**, 1537 (1940).
(86) XeO_2F_2: S. W. Peterson, R. D. Willett, and J. L. Huston, *J. Chem. Phys.* **59**, 453 (1973).
(87) XeO_3F^{-}: D. J. Hodgson and J. A. Ibers. *Inorg. Chem.* **8**, 326 (1969).
(88) $XeOF_3^{+}$: D. E. McKee, C. J. Adams, A. Zalkin, and N. Bartlett, *Chem. Commun.* 26 (1973).
(89) $ClOF_3$: K. O. Christe and E. C. Curtis, *Inorg. Chem.* **11**, 2196 (1972).
(90) G. A. Ozin and A. Vander Voet, *Chem. Commun.* 1489 (1970); *Can. J. Chem.* **49**, 704 (1971); *J. Mol. Struct.* **10**, 397 (1971).
(91) I. R. Beattie *et al.*, *J. Chem. Soc. Dalton* 1747 (1974).
(92) J. R. Morton and K. F. Preston, *J. Chem. Phys.* **58**, 3112 (1973).
(93) S. Durmaz, J. N. Murrell, and J. B. Pedley, *Chem. Commun.* 933 (1972).

AB_6, *12-Electron Complexes*

(94) AlF_6^{3-}: W. Viebahn, *Z. Anorg. Allg. Chem.* **386**, 335 (1971).
(95) GaF_6^{3-}: S. Schwarzmann, *Z. Kristallogr.* **120**, 286 (1964).
(96) InF_6^{3-}: H. Bode and E. Voss, *Z. Anorg. Allg. Chem.* **290**, 1 (1957).
(97) $InCl_6^{3-}$: J. G. Contreras and D. G. Tuck, *Inorg. Chem.* **11**, 2967 (1972); A. K. Dublish and D. K. Sharma, *Spectrosc. Lett.* **5**, 387 (1972).
(98) $InBr_6^{3-}$: J. Gislason, M. H. Lloyd, and D. G. Tuck, *Inorg. Chem.* **10**, 1907 (1971); L. Waterworth and I. J. Worrall, *Inorg. Nucl. Chem. Lett.* **8**, 123 (1972).
(99) TlF_6^{3-}: H. Bode and E. Voss, *Z. Anorg. Allg. Chem.* **290**, 1 (1957).
(100) $TlCl_6^{3-}$, $TlBr_6^{3-}$: T. Watanabe, M. Atoji, and C. Okazaki, *Acta Crystallogr.* **3**, 405 (1950).
(101) SiF_6^{2-}: W. C. Hamilton, *Acta Crystallogr.* **15**, 353 (1962).

(102) GeF_6^{2-}: J. L. Hoard and W. B. Vincent, *J. Am. Chem. Soc.* **61**, 2849 (1939).

(103) $GeCl_6^{2-}$: A. W. Laubengayer, O. B. Billings, and A. E. Newkirk, *J. Am. Chem. Soc.* **62**, 546 (1940).

(104) SnF_6^{2-}: E. A. Marseglia and I. D. Brown, *Acta Crystallogr.* **B29**, 1352 (1973).

(105) $SnCl_6^{2-}$: G. Engel, *Z. Kristallogr.* **90**, 341 (1935).

(106) $SnBr_6^{2-}$: G. Markstein and H. Nowotny, *Z. Kristallogr.* **100**, 265 (1939).

(107) SnI_6^{2-}: W. Werker, *Rec. Trav. Chim.* **58**, 257 (1939).

(108) PbF_6^{2-}: P. Charpin, H. Marquet-Ellis, N. Nghi, and P. Plurien, *C. R. Acad. Sci. Paris C* **275**, 1503 (1972).

(109) $PbCl_6^{2-}$: G. Engel, *Z. Kristallogr.* **90**, 341 (1935).

(110) PF_6^-: R. Destro, T. Pilati, and M. Simonetta, *J. Am. Chem. Soc.* **98**, 1999 (1976); M. Cowie, B. L. Haymore, and J. A. Ibers, *J. Am. Chem. Soc.* **98**, 7608 (1976).

(111) PCl_6^-: D. Clark, H. M. Powell, and A. F. Wells, *J. Chem. Soc.* 642 (1942).

(112) PBr_6^-: G. S. Harris and D. S. Payne, *J. Chem. Soc.* 4617 (1956).

(113) AsF_6^-: F. O. Sladky, P. A. Bulliner, N. Bartlett, B. G. de Boer, and A. Zalkin, *Chem. Commun.* 1048 (1968).

(114) $AsCl_6^-$: I. R. Beattie, T. Gilson, K. Livingston, V. Fawcett, and G. A. Ozin, *J. Chem. Soc. A* 713 (1967).

(115) SbF_6^-: A. J. Edwards and R. J. C. Sills, *J. Chem. Soc. A* 2697 (1970); J. Passmore, P. Taylor, T. K. Whidden, and P. White, *Chem. Commun.* 689 (1976).

(116) $SbCl_6^-$: C. G. Vouk and E. H. Wiebenga, *Acta Crystallogr.* **12**, 859 (1959).

(117) $SbBr_6^-$: S. L. Lawton and R. A. Jacobson, *Inorg. Chem.* **5**, 743 (1966); S. L. Lawton, R. A. Jacobson, and R. S. Frye, *Inorg. Chem.* **10**, 701 (1971).

(118) BiF_6^-: T. Surles, L. A. Quarterman, and H. H. Hyman, *J. Inorg. Nucl. Chem.* **35**, 670 (1973).

(119) SF_6, SeF_6: V. C. Ewing and L. E. Sutton, *Trans. Faraday Soc.* **59**, 1241 (1963).

(120) TeF_6: H. M. Seip and R. Stølevik, *Acta Chem. Scand.* **20**, 1535 (1966).

(121) ClF_6^+: K. O. Christe, *Inorg. Nucl. Chem. Lett.* **8**, 741 (1972); K. O. Christe, J. F. Hon, and D. Pilipovich, *Inorg. Chem.* **12**, 84 (1973); K. O. Christe, *Inorg. Chem.* **12**, 1580 (1973).

(122) BrF_6^+: R. J. Gillespie and G. J. Schrobilgen, *Chem. Commun.* 90 (1974); *Inorg. Chem.* **13**, 1230 (1974).

(123) IF_6^+: F. Seel and O. Detmer, *Z. Anorg. Allg. Chem.* **301**, 113 (1959); K. O. Christe and W. Sawodny, *Inorg. Chem.* **6**, 1783 (1967); **7**, 1685 (1968); K. O. Christe, *Inorg. Chem.* **9**, 2801 (1970); M. Brownstein and H. Selig, *Inorg. Chem.* **11**, 656 (1972).

AB_6, *14-Electron Complexes*

(124) $SnBr_6^{4-}$: J. D. Donaldson, J. Silver, S. Hadjiminolis, and S. D. Ross, *J. Chem. Soc. Dalton* 1500 (1975).

(125) SnI_6^{4-}: Yu. N. Denisov, N. S. Malova, and P. I. Federov, *Russ. J. Inorg. Chem.* **19**, 443 (1974).

(126) PbF_6^{4-}: O. Schmitz-Dumont, G. Bergerhoff, and E. Hartert, *Z. Anorg. Allg. Chem.* **283**, 314 (1956).

(127) $PbCl_6^{4-}$, $PbBr_6^{4-}$: C. K. Møller, *Mat. Fys. Medd. Dan. Vid. Selsk.* **32**, no. 3 (1960).

(128) PbI_6^{4-}: V. A. Federov, N. P. Samsonova, and V. E. Mironov, *Russ. J. Inorg. Chem.* **17**, 674 (1972).

(129) PCl_6^{3-}, $AsCl_6^{3-}$: S. N. Nabi, A. Hussain, and N. N. Ahmed, *J. Chem. Soc. Dalton* 1199 (1974).

(130) SbF_6^{3-}: D. Pavlov and D. Lazarov, *Russ. J. Inorg. Chem.* **3**, 142 (1958).

(131) $SbCl_6^{3-}$: D. R. Schroeder and R. A. Jackson, *Inorg. Chem.* **12**, 210 (1973).

(132) $SbBr_6^{3-}$: S. L. Lawton and R. A. Jacobson, *Inorg. Chem.* **5**, 743 (1966); S. L. Lawton, R. A. Jacobson, and R. S. Frye, *Inorg. Chem.* **10**, 701 (1971).

(133) SbI_6^{3-}: M. A. Hooper and D. W. James, *Aust. J. Chem.* **26**, 1401 (1973).

(134) BiF_6^{3-}: A. Cousson, A. Vedrine, and J.-C. Cousseins, *C. R. Acad. Sci. Paris C* **247**, 864 (1973).

(135) $BiCl_6^{3-}$: L. R. Morss and W. R. Robinson, *Acta Crystallogr.* **B28**, 653 (1972).

(136) $BiBr_6^{3-}$: W. G. McPherson and E. A. Meyers, *J. Phys. Chem.* **72**, 3117 (1968).

(137) BiI_6^{3-}: R. A. Spragg, H. Stammreich, and Y. Kawano, *J. Mol. Struct.* **3**, 305 (1969).

(138) $SeCl_6^{2-}$: G. Engel, *Z. Kristallogr.* **90**, 341 (1935).

(139) $SeBr_6^{2-}$: J. L. Hoard and B. N. Dickinson, *Z. Kristallogr.* **84**, 436 (1932).

(140) SeI_6^{2-}: P. J. Hendra and Z. Jović, *J. Chem. Soc. A* 600 (1968).

(141) TeF_6^{2-}: E. E. Aynsley and G. Hetherington, *J. Chem. Soc.* 2802 (1953).

(142) $TeCl_6^{2-}$: A. C. Hazell, *Acta Chem. Scand.* **20**, 165 (1966).

(143) $TeBr_6^{2-}$: A. K. Das and I. D. Brown, *Can. J. Chem.* **44**, 939 (1966).

(144) TeI_6^{2-}, $PoCl_6^{2-}$, $PoBr_6^{2-}$, PoI_6^{2-}: K. W. Bagnall, R. W. M. D'Eye, and J. H. Freeman, *J. Chem. Soc.* 3385 (1956).

(145) ClF_6^{-}: J. P. Faust, A. W. Jache, and A. J. Klanica, *Chem. Abstr.* **76**, 47891a (1972); **79**, 55557y (1973).

(146) BrF_6^{-}: E. D. Whitney, R. O. MacLaren, C. E. Fogle, and T. J. Hurley, *J. Am. Chem. Soc.* **86**, 2583 (1964); H. Meinert and U. Gross, *Z. Chem.* **11**, 469 (1971).

(147) IF_6^{-}: K. O. Christe, J. P. Guertin, and W. Sawodny, *Inorg. Chem.* **7**, 626 (1968); K. O. Christe, *Inorg. Chem.* **11**, 1215 (1972).

(148) XeF_6: R. M. Gavin and L. S. Bartell, *J. Chem. Phys.* **48**, 2460 (1968).

AB_6, 13-Electron Radicals

(149) ClF_6, BrF_6, IF_6: A. R. Boate, J. R. Morton, and K. F. Preston, *Inorg. Chem.* **14**, 3127 (1975); *J. Phys. Chem.* **80**, 2954 (1976); K. Nishikida, F. Williams, G. Mamantov, and N. Smyrl, *J. Am. Chem. Soc.* **97**, 3526 (1975); *J. Chem Phys.* **63**, 1693 (1975).

(150) SF_6^{-}, SeF_6^{-}, TeF_6^{-}: J. R. Morton, K. F. Preston, and J. C. Tait, *J. Chem. Phys.* **62**, 2029 (1975).

(151) AsF_6^{2-}, SbF_6^{2-}: A. R. Boate, J. R. Morton, and K. F. Preston, *Chem. Phys. Lett.* **50**, 65 (1977).

AB_6 Oxides

(152) SbO_6^{7-}: H. Siebert, *Z. Anorg. Allg. Chem.* **301**, 161 (1959).

(153) TeO_6^{6-}: H. T. Evans, Jr., *Acta Crystallogr.* **B30**, 2095 (1974).

(154) IO_6^{5-}: L. Helmholz, *J. Am. Chem. Soc.* **59**, 2036 (1937); Y. D. Feikema, *Acta Crystallogr.* **20**, 765 (1966).

(155) XeO_6^{4-}: W. C. Hamilton, J. A. Ibers, and D. R. Mackenzie, *Science* **141**, 532 (1963); A. Zalkin, J. D. Forrester, D. H. Templeton, S. M. Williamson, and C. W. Koch, *Science* **142**, 501 (1963).

General References

(156) J. Easterfield and J. W. Linnett, *Chem. Commun.* 64 (1970).

(157) R. Clampitt and D. K. Jeffries, *Nature (London)*, **226**, 141 (1970).

(158) K. Hiroaka and P. Kebarle, *J. Chem. Phys.* **62**, 2267 (1975).

(159) J. P. Lowe, *J. Am. Chem. Soc.* **92**, 3799 (1970); *Science* **179**, 527 (1973).

(160) L. S. Bartell and R. M. Gavin, Jr., *J. Chem. Phys.* **48**, 2466 (1968).
(161) K. S. Pitzer and L. S. Bernstein, *J. Chem. Phys.* **63**, 3849 (1975).
(162) P. Labonville, J. R. Ferraro, M. C. Wall, and L. J. Basile, *Coord. Chem. Rev.* **7**, 257 (1972).
(163) T. Barrowcliffe, I. R. Beattie, P. Day, and K. Livingston, *J. Chem. Soc. A* 1810 (1967).

AB_5, Ten-Electron Complexes

(164) AlF_5^{2-}, GaF_5^{2-}, InF_5^{2-}: P. Bukovec, B. Orel, and J. Siftar, *Monatsh. Chem.* **102**, 885 (1971); **104**, 194 (1973).
(165) $InCl_5^{2-}$, $TlCl_5^{2-}$: G. Joy, A. P. Gaughan, Jr., I. Wharf, D. F. Shriver, and J. P. Dougherty, *Inorg. Chem.* **14**, 1795 (1975).
(166) SiF_5^-: H. C. Clark and K. R. Dixon, *Chem. Commun.* 717 (1967).
(167) $SiCl_5^-$: I. R. Beattie and K. M. Livingston, *J. Chem. Soc. A* 859 (1969).
(168) GeF_5^-: (a) I. Wharf and M. Onyszchuk, *Can. J. Chem.* **48**, 2250 (1970); (b) K. O. Christe, R. D. Wilson, and I. B. Goldberg, *Inorg. Chem.* **15**, 1271 (1976).
(169) $GeCl_5^-$: I. R. Beattie, T. Gilson, K. Livingston, V. Fawcett, and G. A. Ozin, *J. Chem. Soc. A* 712 (1967).
(170) $SnCl_5^-$: R. F. Bryan, *J. Am. Chem. Soc.* **86**, 733 (1964).
(171) $SnBr_5^-$: J. A. Creighton and J. H. S. Green, *J. Chem. Soc. A* 808 (1968).
(172) PF_5: K. W. Hansen and L. S. Bartell, *Inorg. Chem.* **4**, 1775 (1965).
(173) PCl_5: (a) W. J. Adams and L. S. Bartell, *J. Mol. Struct.* **8**, 23 (1971); (b) R. W. Suter, H. C. Knachel, V. P. Petro, J. H. Howatson, and S. G. Shore, *J. Am. Chem. Soc.* **95**, 1474 (1973).
(174) PBr_5: W. Gabes and K. Olie, *Acta Crystallogr.* **B26**, 443 (1970).
(175) AsF_5: F. B. Clippard, Jr., and L. S. Bartell, *Inorg. Chem.* **9**, 805 (1970).
(176) SbF_5: J. Gaunt and J. B. Ainscough, *Spectrochim. Acta* **10**, 57 (1957); L. E. Alexander and I. R. Beattie, *J. Chem. Phys.* **56**, 5829 (1972).
(177) $SbCl_5$: (a) S. M. Ohlberg, *J. Am. Chem. Soc.* **81**, 811 (1959); (b) G. L. Carlson, *Spectrochim. Acta* **19**, 1291 (1963).
(178) BiF_5: C. Hebecker, *Z. Anorg. Allg. Chem.* **384**, 111 (1971).
(179) $InCl_5^{2-}$: D. S. Brown, F. W. F. Einstein, and D. G. Tuck, *Inorg. Chem.* **8**, 14 (1969); D. F. Shriver and I. Wharf, *Inorg. Chem.* **8**, 2167 (1969).
(180) $TlCl_5^{2-}$: D. F. Shriver and I. Wharf, *Inorg. Chem.* **8**, 2167 (1969).
(181) R. S. Berry, *J. Chem. Phys.* **32**, 933 (1960).
(182) F. Klanberg and E. L. Muetterties, *Inorg. Chem.* **7**, 155 (1968).
(183) I. Ugi, D. Marquarding, H. Klusacek, P. Gillespie, and F. Ramirez, *Accounts Chem. Res.* **4**, 288 (1971).

AB_5, Ten-Electron Mixed-Ligand Complexes

(184) PF_nCl_{5-n}: J. E. Griffiths, R. P. Carter, Jr., and R. R. Holmes, *J. Chem. Phys.* **41**, 863 (1964).
(185) ClO_2F_3: K. O. Christe, *Inorg. Nucl. Chem. Lett.* **8**, 457 (1972); K. O. Christe and R. D. Wilson, *Inorg. Chem.* **12**, 1356 (1973); K. O. Christe and E. C. Curtis, *Inorg. Chem.* **12**, 2245 (1973).
(186) IO_2F_3: I. R. Beattie and G. J. Van Schalkwyk, *Inorg. Nucl. Chem. Lett.* **10**, 343 (1974).
(187) SOF_4: G. Gundersen and K. Hedberg, *J. Chem. Phys.* **51**, 2500 (1969).

AB_5, 12-Electron Complexes

(188) $PbBr_5^{3-}$: F. Vierling, *Bull. Soc. Chim. Fr.* 2563 (1972); A. G. Galinos and I. Triantafillopoulou, *Isr. J. Chem.* **12**, 771 (1974).
(189) SbF_5^{2-}: R. R. Ryan and D. T. Cromer, *Inorg. Chem.* **11**, 2322 (1972).
(190) $SbCl_5^{2-}$: M. Webster and S. Keats, *J. Chem. Soc. A* 298 (1971).
(191) $SbBr_5^{2-}$: H. A. Abdel-Rehim and E. A. Meyers, *Crystallogr. Struct. Commun.* **2**, 45 (1973).
(192) SbI_5^{2-}: M. A. Hooper and D. W. James, *J. Inorg. Nucl. Chem.* **35**, 2335 (1973).
(193) $BiCl_5^{2-}$: R. P. Oertel and R. A. Plane, *Inorg. Chem.* **6**, 1960 (1967); R. A. Walton, *Spectrochim. Acta* **24A**, 1527 (1968).
(194) $BiBr_5^{2-}$: W. G. McPherson and E. A. Meyers, *J. Phys. Chem.* **72**, 532 (1968).
(195) BiI_5^{2-}: M. A. Hooper and D. W. James, *J. Inorg. Nucl. Chem.* **35**, 2335 (1973).
(196) SF_5^-: L. F. Drullinger and J. E. Griffiths, *Spectrochim. Acta* **27A**, 1793 (1971).
(197) SeF_5^-: K. O. Christe, E. C. Curtis, C. J. Schack, and D. Pilipovich, *Inorg. Chem.* **11**, 1679 (1972).
(198) $SeCl_5^-$: H. Gerding and J. C. Duinker, *Rev. Chim. Miner.* **3**, 815 (1966).
(199) $SeBr_5^-$: P. J. Hendra and Z. Jovic, *J. Chem. Soc. A* 600 (1968).
(200) TeF_5^-: S. H. Mastin, R. R. Ryan, and L. B. Asprey, *Inorg. Chem.* **9**, 2100 (1970).
(201) $TeCl_5^-$: E. E. Aynsley and A. C. Hazell, *Chem. Ind. (London)* 611 (1963).
(202) $TeBr_5^-$: G. A. Ozin and A. Vander Voet, *J. Mol. Struct.* **13**, 435 (1972).
(203) PoI_5^-: K. W. Bagnall, R. W. M. D'Eye, and J. H. Freeman, *J. Chem. Soc.* 3385 (1956).
(204) ClF_5: G. M. Begun, W. H. Fletcher, and D. F. Smith, *J. Chem. Phys.* **42**, 2236 (1965).
(205) BrF_5, IF_5: A. G. Robiette, R. H. Bradley, and P. N. Brier, *Chem. Commun.* 1567 (1971).
(206) XeF_5^+: N. Bartlett, M. Gennis, D. D. Gibler, B. K. Morrell, and A. Zalkin, *Inorg. Chem.* **12**, 1717 (1973).

AB_5, 12-Electron Mixed-Ligand Complexes

(207) $TeOF_4^{2-}$: J. B. Milne and D. Moffett, *Inorg. Chem.* **12**, 2240 (1973).
(208) $ClOF_4^-$: K. O. Christe and E. C. Curtis, *Inorg. Chem.* **11**, 2209 (1972).
(209) $BrOF_4^-$: R. J. Gillespie and P. Spekkens, *J. Chem. Soc. Dalton* 2391 (1976).
(210) IOF_4^-: R. R. Ryan and L. B. Asprey, *Acta Crystallogr.* **B28**, 979 (1972).
(211) $XeOF_4$: E. J. Jacob, H. B. Thompson, and L. S. Bartell, *J. Mol. Struct.* **8**, 383 (1971).

AB_5, 11-Electron Radicals

(212) SF_5: J. R. Morton and K. F. Preston, *Chem. Phys. Lett.* **18**, 98 (1973); J. Gawlowski and J. A. Herman, *Can. J. Chem.* **52**, 3631 (1974).
(213) PF_5^-: S. P. Mishra and M. C. R. Symons, *Chem. Commun.* 279 (1974).
(214) PCl_5^-: S. P. Mishra and M. C. R. Symons, *J. Chem. Soc. Dalton* 139 (1976).

AB_5, Semiempirical and Ab Initio Calculations

(215) A. Rauk, L. C. Allen, and K. Mislow, *J. Am. Chem. Soc.* **94**, 3035 (1972).
(216) R. Hoffmann, J. M. Howell, and E. L. Muetterties, *J. Am. Chem. Soc.* **94**, 3047 (1972).
(217) A. Strich and A. Veillard, *J. Am. Chem. Soc.* **95**, 5574 (1973).
(218) R. D. Brown and J. B. Peel, *Aust. J. Chem.* **21**, 2605 (1968).
(219) P. C. Van Der Voorn and R. S. Drago, *J. Am. Chem. Soc.* **88**, 3255 (1966).
(220) D. P. Santry and G. A. Segal, *J. Chem. Phys.* **47**, 158 (1967).

(221) R. S. Berry, M. Tamres, C. J. Ballhausen, and H. Johansen, *Acta Chem. Scand.* **22**, 231 (1968).
(222) J. B. Florey and L. C. Cusachs, *J. Am. Chem. Soc.* **94**, 3040 (1972).
(223) J. M. Howell, J. R. Van Wazer, and A. R. Rossi, *Inorg. Chem.* **13**, 1746 (1974).
(224) J. M. Howell, *J. Am. Chem. Soc.* **97**, 3930 (1975).
(225) A. R. Gregory, *Chem. Phys. Lett.* **28**, 552 (1974).
(226) M. F. Guest, M. B. Hall, and I. H. Hillier, *J. Chem. Soc. Faraday II* **69**, 1829 (1973).

General References

(227) L. S. Bernstein, S. Abramowitz, and I. W. Levin, *J. Chem. Phys.* **64**, 3228 (1976).
(228) R. R. Holmes and R. M. Deiters, *Inorg. Chem.* **7**, 2229 (1968); R. R. Holmes, R. M. Deiters, and J. A. Golen, *Inorg. Chem.* **8**, 2612 (1969); R. R. Holmes, *Accounts Chem. Res.* **5**, 296 (1972).
(229) W. G. Klemperer, J. K. Krieger, M. D. McCreary, E. L. Muetterties, D. D. Traficante, and G. M. Whitesides, *J. Am. Chem. Soc.* **97**, 7023 (1975).
(230) S. Ahrland and L. Grenthe, *Acta Chem. Scand.* **11**, 1111 (1957).
(231) R. E. Rundle, *J. Am. Chem. Soc.* **85**, 112 (1963).
(232) D. C. Frost, C. A. McDowell, J. S. Sandhu, and D. A. Vroom, *J. Chem. Phys.* **46**, 2008 (1967).
(233) F. C. Fehsenfeld, *J. Chem. Phys.* **54**, 438 (1971).
(234) J. I. Musher, *Angew. Chem., Int. Ed. Engl.* **8**, 54 (1969).
(235) S. Reichman and F. Schreiner, *J. Chem. Phys.* **51**, 2355 (1969).
(236) D. E. McKee, A. Zalkin, and N. Bartlett, *Inorg. Chem.* **12**, 1713 (1973).
(237) D. W. Magnuson, *J. Chem. Phys.* **27**, 223 (1957); R. D. Burbank and F. N. Bensey, *J. Chem. Phys.* **27**, 982 (1957).
(238) N. N. Greenwood, B. P. Straughan, and A. E. Wilson, *J. Chem. Soc. A* 1479 (1966).
(239) L. Kolditz, *Adv. Inorg. Chem. Radiochem.* **7**, 1 (1965).
(240) R. D. Burbank and G. R. Jones, *Science* **168**, 248 (1970).
(241) H. H. Hyman and L. A. Quarterman, *in* "Noble-Gas Compounds" (H. H. Hyman, ed.), p. 275. Univ. of Chicago Press, Chicago, Illinois, 1963.
(242) N. Bartlett and F. O. Sladky, *J. Am. Chem. Soc.* **90**, 5316 (1968).
(243) G. A. Ozin and A. Vander Voet, *Chem. Commun.* 896 (1970).
(244) D. M. Yost and J. B. Hatcher, *J. Am. Chem. Soc.* **53**, 2549 (1931).
(245) N. Katsaros and J. W. George, *Inorg. Chem.* **8**, 759 (1969).
(246) B. Cohen and R. D. Peacock, *Adv. Fluorine Chem.* **6**, 343 (1970).
(247) E. E. Aynsley, R. D. Peacock, and P. L. Robinson, *J. Chem. Soc.* 1231 (1952); F. Brown and P. L. Robinson, *J. Chem. Soc.* 3147 (1955).
(248) F. Nyman and H. L. Roberts, *J. Chem. Soc.* 3180 (1962).
(249) C. W. Tullock, D. D. Coffman, and E. L. Muetterties, *J. Am. Chem. Soc.* **86**, 357 (1964).

5 HAB and H₂AB

HAB

Table I lists some known and proposed molecules of the class HAB (*1–42*). With only three atoms these molecules are either linear or bent. Not included in Table I are the AHB species such as the 16-electron ions FHF^-, $ClHCl^-$, $BrHBr^-$, and IHI^- that are linear and symmetric, with the hydrogen midway between the two halogens (*43,44*). The occupied valence MOs for linear HAB molecules can be derived from those for the diatomic molecule AB by combining a hydrogen $1s$ AO in-phase with the σ MOs of AB. Nodal planes prevent any hydrogen $1s$ contribution to the π orbitals. Figure 1 shows schematic MO pictures correlating orbitals for linear and bent HAB. Changes in orbital energy with angular variations are easy to interpret from AO overlap changes, with the exceptions of 3σ-$3a'$ and $1\pi_x$-$4a'$. Both 3σ and $1\pi_x$ must yield orbitals of a' symmetry in bent geometry. Intended orbital correlations are indicated by dashed arrows in Fig. 1. The crossing of two a' orbitals is forbidden by the noncrossing rule. Instead, these two orbitals mix and diverge as shown by the solid arrows for 3σ-$3a'$ and $1\pi_x$-$4a'$.

Molecules with nine or ten valence electrons, such as HBN or HCN, are linear, the highest occupied MO being $1\pi_x$-$4a'$. Although the ground states of HCN and HCP are linear, they both have bent excited states. Excitation transfers an electron from $1\pi_x$-$4a'$, the orbital that holds the molecule linear, to $2\pi_x$-$5a'$, an orbital that goes to significantly lower energy in the bent shape. Thus, ten-electron molecules should have linear ground states and bent

113

TABLE I

Shapes of known or proposed HAB molecules

Number of valence electrons	HAB	Shape
9	HCC, HCN⁺, HNB	
10	HNN⁺, HCN, HNC, HCP, HBO, HBS	linear
11	HCO, HNN, HCN⁻, HNO⁺, HON⁺, HBO⁻, HBF	
12	HCF, HCCl, HNO, HON, HPO, HSiCl, HSiBr, HSiI	
13	HOO, HSS, HSO, HNF	bent
14	HOF, HOCl, HOBr	

excited states. The 11-electron radical HCO has a bent ground state with one electron in $5a'$. Excitation moves this electron from $5a'$, the orbital that makes ground state HCO bent, and adds it to $2\pi_y$-$2a''$, an orbital with no geometry preference. Lower occupied orbitals, the same ones that make the HCN ground state linear, give excited HCO linear geometry. Molecules with 12, 13, or 14 electrons should be bent in both ground and excited states because at least one electron occupies $5a'$.

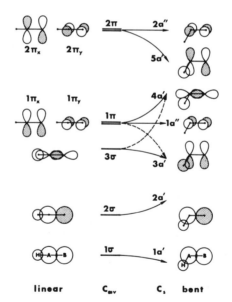

FIG. 1 Qualitative MO correlation diagram for linear and bent HAB molecules. Dashed arrows indicate intended correlations that are blocked by the noncrossing rule. [Reprinted with permission from Gimarc, *J. Am. Chem. Soc.* **93**, 815 (1971). Copyright by the American Chemical Society.]

The $1\pi_x$ and $2\pi_x$ MOs are bonding and antibonding combinations, respectively, of parallel p AOs. Suppose that the terminal B atom in HAB is more electronegative than the central atom A. Then perturbation theory would give $1\pi_x$ a larger contribution from p_B than from p_A because p_B has the lower energy. The composition of $2\pi_x$ would be just the opposite, with the higher energy p_A AO having the larger coefficient (**1**). The related $3a'$ and $5a'$ MOs of bent geometry might maintain a similar balance of p AO

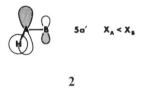

1

contributions, although possibilities for MO mixing with nearby orbitals of similar symmetry in bent geometry weaken the analogy. But assuming that $5a'$ can be adequately represented, as in **2**, and that a larger p AO on A favors smaller HAB angles through improved AO overlap, we can predict that in

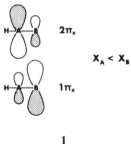

2

molecules with 11 electrons or more as well as excited states of 10-electron molecules, HAB angles will be smaller for a larger electronegativity difference ΔX between A and B such that the terminal B is more electronegative. Table II contains a number of HAB bond angles, showing that this rule usually holds although there are some failures. For example, the angle in HOF($14e$) is smaller than that in HOCl, the angle in HPO($12e$) is smaller than that in HNO, and excited HCN($10e$) is more sharply bent than excited HCP. On the other hand, bond angles are nearly constant through the series HSiCl, HSiBr, and HSiI, although the electronegativity rule predicts decreasing bond angles. Ab initio calculations give $HON^+(11e)$ a smaller angle than for HNO^+, just the reverse of the qualitative model prediction.

TABLE II

HAB bond angles

Valence electrons	Molecule (angle), increasing $\Delta X \rightarrow$
10^a	HCP(130°), HCN(125°)
11	HNN(118°, 124°), HCO(127.5°), HBF(121°) HON⁺(123.7°), HNO⁺(130°)
12	HCCl(103.4°), HCF(101.6°) HNO(108.6°), HPO(104.7°) HSiI(102.7°), HSiBr(102.9°), HSiCl(102.8°)
13	HOO(104.6°), HNF(101°) HSO(100.6°)
14	HOCl(102.5°), HOF(96.8°)

^{a} Excited state geometry.

HYDROGEN BONDING

Unlike the other HAB species, the 16-electron ions such as FHF^- are linear and symmetric with the hydrogen on the axis and midway between the two halogen atoms. One can think of an ordinary HAB molecule (14 electrons or less) as an AB^- ion with a proton attached to one end. For example, HCN, HNO, and HOCl are all weak acids and the corresponding AB^- ions are known. One would not expect the ion F_2^{2-} (16 electrons, isoelectronic with Ne_2) to be bound because the stabilization resulting from four occupied bonding valence MOs is more than canceled by the effect of an equal number of filled antibonding orbitals. Therefore, an ion such as HFF^- should be unstable relative to HF and F^-. For the FHF^- system, the insertion of a hydrogen with its $1s$ AO between the two halogens stabilizes the antibonding σ_u and π_g orbitals (particularly the highest occupied $2\sigma_u$ MO that has sigma-type p AOs pointing directly at each other out-of-phase) by separating the antibonding or out-of-phase fragments, thereby lowering their out-of-phase overlaps and making them nonbonding. At the same time, the central hydrogen $1s$ orbital provides in-phase overlaps to maintain the bonding character of the σ_g MOs. Figure 2 compares MOs for X_2^{2-} and XHX^-. Only the bonding π_u orbitals are raised in energy (to become nonbonding) by the increased separation of the two fluorine atoms. In the π_u MOs the hydrogen is on the nodal surface and its $1s$ AO cannot participate in any overlap stabilization. The result is a change from four bonding and four antibonding MOs for X_2^{2-} to two bonding and six nonbonding MOs for XHX^-.

Preparation of the neutral radicals XHX (X = Cl, Br, and I) has been reported (46) and refuted (47, 48). If the $2\sigma_u$ MO of XHX^- is truly non-

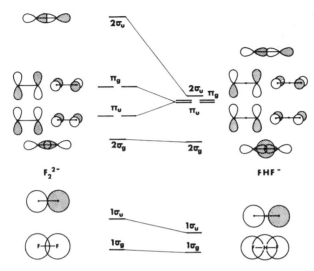

FIG. 2   Relative energy levels and MO pictures for F_2^{2-} and FHF^-. [Reprinted with permission from Gimarc, *J. Am. Chem. Soc.* **93**, 815 (1971). Copyright by the American Chemical Society.]

bonding, then one would expect the XHX radicals to have hydrogen bond energies that are comparable to those of the respective XHX⁻ anions. If $2\sigma_u$ were more like the antibonding orbital of X_2^{2-}, then the XHX radical might have a stronger bond than the anion. The failure of attempts to produce the radicals seems to imply a failure of the qualitative model. Furthermore, ab initio calculations for the radicals FHF (*49*) and ClHCl (*50*) yield asymmetrical structures XH---X that are only slightly more stable than separated XH + X. In particular, the calculated hydrogen bond energy of FHF is only 3 kcal/mole compared to the experimental value of 37 kcal/mole for FHF⁻ (*51*). The hydrogen bond energy of the anion ClHCl⁻ is 19 kcal/mole (*50*), but the ab initio calculated hydrogen bond energy for the radical ClHCl is 7 to 15 kcal/mole. Thus, the ab initio calculations suggest that both radicals are considerably less stable than the qualitative model predicts them to be.

It has been proposed that FHeF might be stable, from analogy with the isoelectronic ion FHF⁻ and FXeF (*52*). However, ab initio valence bond (*53*) and SCF MO (*49*) calculations indicate that while FHF⁻ is stable, FHeF should be unstable. The qualitative model can rationalize those conclusions. Hydrogen bonding results from antibonding or out-of-phase pieces of MOs being separated far enough to make all antibonding orbitals nonbonding, while some of the bonding orbitals retain their bonding character through new overlaps provided by the inserted hydrogen 1s AO. The larger charge of the He nucleus makes a He 1s AO considerably smaller than a hydrogen 1s. In order to maintain effective in-phase overlaps in the two σ_g MOs of FHeF, a He 1s orbital cannot separate the F atoms far enough to stabilize significantly the antibonding MOs. Therefore, net bonding in FHeF is not likely.

The same qualitative MO model of hydrogen bonding can be used to describe the rough features of the water dimer $(H_2O)_2$, also a 16-electron system. Structures **a–d** in **3** are some possible shapes for the water dimer.

3

The occupied MOs of $(H_2O)_2$ will be determined primarily by the valence AOs of the two oxygens. The most stable shape of the dimer will be that which gives maximum separation of the O atoms in order to stabilize the antibonding MOs while maintaining, through overlaps with a hydrogen 1s AO, the bonding character of at least some MOs. Of the possibilities shown above, structure **a** is obviously the poorest, **b** and **c** are better, but **d** is the best. Similar conclusions have been reached on the basis of ab initio calculations (54). The qualitative model also predicts a linear X-H-Y bond in other 16-electron dimers or pairs such as $(HF)_2$, H_3N-HF, and H_2O-HF. These same arguments can be extended to account for linear hydrogen bonds in polymers such as $(H_2O)_n$ and $(HF)_n$. The $4n$ MOs of lowest energy are those made from the $4n$ AOs of F or O. These $4n$ MOs are filled with $8n$ electrons, equal numbers of bonding and antibonding MOs being occupied. (In some cases there may also be pairs of nonbonding MOs.) Again, one would not expect structures with directly bonded O-O or F-F combinations. Hydrogens with their 1s AOs positioned on the O---O or F---F axes would turn antibonding MOs into nonbonding MOs while preserving some bonding combinations leading to stable chains or cyclic structures (55).

A note of caution is in order here. Allen has pointed out that the hydrogen bonding is largely electrostatic (56). Qualitative MO theory does not satisfactorily account for electrostatic interactions that, therefore, must be introduced separately. In the hydrogen bonding examples discussed here, electrostatics and simple MO arguments usually lead to the same conclusions, thereby reinforcing each other, a situation that may not always occur.

OTHER POSSIBLE AHA$^\pm$ SYSTEMS

Ab initio calculations indicate that the two-valence electron ion Li$_2$H$^+$ is stable and has the linear symmetric structure LiHLi$^+$ (57,58). The qualitative MO model predicts this ion to be bent or triangular like the isoelectronic species LiH$_2$$^+$ and H$_3$$^+$. An explanation of the linear LiHLi$^+$ structure must rest on the peculiar properties of the Li-Li and Li-H bonds. The Li$_2$ bond is the longest (2.67 Å) and weakest (26 kcal/mole) of all the first-row homonuclear diatomic molecules. The Li-Li bond is almost twice as long as the bond in Li-H, the ratio $R_{\text{Li-H}}/R_{\text{Li-Li}}$ being around 0.6. With very little difficulty a proton could slip between two lithium atoms in Li$_2$. Because of its small effective nuclear charge, the lithium 2s orbital is quite large. The hydrogen 1s orbital can overlap effectively with both Li atom 2s orbitals without disturbing the normal 2s, 2s overlap. In this way the proton could form two Li-H bonds and still maintain the Li-Li bond. The ab initio results predict that Li$_2$H$^+$ is more stable than LiH and Li$^+$ by 60 to 63 kcal/mole, an amount only slightly larger than the energy of the LiH bond (58 kcal/mole).

Ab initio calculations for Li$_2$H$^-$ and Be$_2$H$^+$ suggest that these four-electron ions should have linear, symmetric structures AHA$^\pm$ and should be stable by about 7 kcal/mole relative to LiH + Li$^-$ and BeH$^+$ + Be, respectively (59,60). The story here must be similar to the one for FHF$^-$. Simple MO theory predicts that Li$_2^{2-}$ and Be$_2$ would not be bound. (Once more, look at the X$_2$ side of Fig. 2.) Insertion of a proton between the two Li or Be atoms would stabilize the antibonding 1σ_u MO by separating two out-of-phase 2s AOs while providing some measure of in-phase overlap among Li and Be atom 2s and hydrogen 1s AOs in 1σ_g to hold together the ions LiHLi$^-$ and BeHBe$^+$.

Extending these arguments one would expect AHA$^\pm$ ions with six-valence electrons or less to be stable with respect to H$^+$ and A$_2$ or A$_2^{2-}$. Any electrons more than six would occupy the destablized π_u MO in AHA$^\pm$. The only information available for comparison is an ab initio calculation that shows a higher energy for BeHBe$^-$ than for BeH + Be. Unfortunately, these are not convenient dissociation products for comparison for the simple MO theory. BeHBe$^-$ might still be metastable.

H$_2$AB

Molecules with the general formula H$_2$AB, in which both hydrogens are bound to a central atom A, are either planar or pyramidal. Some known or proposed examples are listed in Table III (61–80). Ab initio MO calculations predict planar geometry for vinylidine carbene H$_2$CC (ten electrons).

TABLE III

Known and proposed H₂AB molecules

Number of valence electrons	H₂AB	Shape
10	H_2CC	
11	H_2CN, H_2CO^+, H_2BO	planar
12	H_2NN, H_2CO, H_2CS, H_2CF^+, H_2BF	
13	H_2NO, H_2CO^-, H_2CF, H_2CCl, H_2CBr, H_2SiF	planar or pyramidal
14	H_2NF, H_2NCl	pyramidal

The ESR spectrum of the methylene imino radical H_2CN (11 electrons) indicates planar geometry. Formaldehyde H_2CO (12 electrons) has a planar ground state but a pyramidal or nonplanar excited state. The 13-electron radical H_2SiF is pyramidal but the isoelectronic radicals H_2CCl and H_2CBr are apparently planar. Chloramine H_2NCl (14 electrons) is pyramidal.

Figure 3 contains schematic pictures of the lower energy valence MOs for H_2AB in planar and pyramidal shapes. The MOs for planar H_2AB are easily formed from A_2 or AB diatomic MOs by adding the two hydrogen $1s$ orbitals in-phase with the AOs on the A atom. The hydrogen $1s$ orbitals

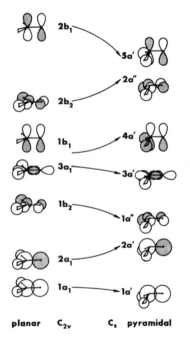

FIG. 3 Qualitative MO correlation diagram for planar and pyramidal H_2AB molecules. [Reprinted with permission from Gimarc, *J. Am. Chem. Soc.* **93**, 815 (1971). Copyright by the American Chemical Society.]

planar   C₂ᵥ   Cₛ pyramidal

mix with the diatomic MOs to form $1a_1$, $2a_1$, and $3a_1$ in Fig. 3. Hydrogen $1s$ AOs mix with the π_x orbitals of AB to form MOs of b_2 symmetry. The π_y diatomic MOs become, unchanged, the b_1 orbitals of planar H_2AB, because the hydrogens lie on the nodal surfaces of these MOs. Thus, the degeneracies of both the bonding and the antibonding π levels are removed. There is also some rearrangement of the energy order of the MOs on going from AB to H_2AB, but this can be understood in terms of AO overlaps and has no bearing on molecular shapes. In the linear HAB system, 3σ was slightly lower in energy than the 1π pair, but in H_2AB the related $3a_1$ orbital lies between $1b_1$ and $1b_2$, which are related to 1π (HAB). Assuming an HAH angle in H_2AB of around $120°$, each hydrogen $1s$ in $3a_1$ is in about 50% of the maximum possible overlap with the p_z orbital on A. This is roughly comparable with the amount of overlap between the p_z and the single hydrogen $1s$ AO in the related 3σ orbital of linear HAB. Therefore, if 3σ is below 1π for linear HAB, then $3a_1$ should be below $1b_1$ for planar H_2AB. In $1b_2$ each hydrogen $1s$ is in about 87% of the maximum possible overlap with the p_x orbital on A. This large $1s$, p_x overlap is presumably enough to place $1b_2$ below $3a_1$ as shown in Fig. 3.

As the H_2AB molecule is folded from planar to pyramidal shape, holding A and B fixed, most of the orbital energy changes follow AO overlap changes. The exceptions are $3a_1$-$3a'$ and $1b_1$-$4a'$. In planar geometry $3a_1$ and $1b_1$ become MOs of a' symmetry and, following the overlap changes, they would converge on each other and possibly intersect. The noncrossing rule prevents this, however; $3a'$ and $4a'$ actually mix and diverge, as shown in Fig. 3. The mixing-produced destabilization of the upper orbital is always greater than the stabilization of the lower orbital, and the net result is an increase in energy for the combined $3a_1$-$3a'$ and $1b_1$-$4a'$ systems on pyramidalization. Notice the changes in overlap between the hydrogen $1s$ and B atom s AOs in $1a_1$-$1a'$ and $2a_1$-$2a'$. In general, an increase in out-of-phase overlap produces an energy increase that is greater than the energy lowering from a similar in-phase overlap change. Two other MOs with comparable in-phase and out-of-phase overlap changes are $1b_2$-$1a''$ and $2b_2$-$2a''$. The orbitals $2b_2$-$2a''$ and below show a net energy increase on pyramidalization.

In formaldehyde (12 electrons) the highest occupied MO in the ground state is $2b_2$-$2a''$ and the molecule is planar. Electron excitation removes an electron from an orbital ($2b_2$-$2a''$) that helps hold H_2CO planar and adds it to an orbital ($2b_1$-$5a'$) that strongly favors the pyramidal structure. Therefore, the excited state of H_2CO should be pyramidal, as observed. An ab initio calculation (73) for the $(2b_2)^2$ singlet configuration of H_2NN (12 electrons) shows the expected planar structure. A matching calculation for the $(2b_2)^1(2b_1)^1$ H_2NN triplet gives a nonplanar structure in accord with the properties of $5a'$. The 13-electron radical H_2CF, with $2b_2$-$2a''$ doubly occupied and one electron in $2b_1$-$5a'$, should also be pyramidal as ab initio calculations suggest (78). Chloramine H_2NCl is pyramidal with $2b_1$-$5a'$ doubly occupied.

Consider the two b_2 MOs of planar geometry. For B more electronegative than A, perturbation arguments give p_B the larger coefficient in $1b_2$ and p_A the larger coefficient in $2b_2$ (4). Since $2b_2$ is occupied in 12-electron molecules, the larger interaction between hydrogen $1s$ and p_A should favor wider HAH angles for more electronegative terminals B. Structural information for the 12-electron series is sketchy but it does lend some support for this rule. The HAH angles determined by microwave spectroscopy for H_2CO and H_2CS are practically the same (116.4° and 116.9°, respectively). An ab initio calculation (75) for H_2CF^+ gives an HCH

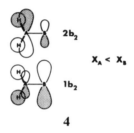

$$2b_2$$
$$X_A < X_B$$
$$1b_2$$

4

angle of 120° ± 2°, greater than those in H_2CO and H_2CS as predicted by the rule. An ab initio SCF MO + CI calculation for planar H_2NN yields an HNH angle of 112°. The qualitative MO model predicts a smaller HAH angle for H_2NN, in which A and B have the same electronegativity, than for H_2CO, for which $X_A < X_B$.

The bonding and nonbonding pi MOs of planar H_2AB are $1b_1$ and $2b_1$, respectively. Again, assuming B to be more electronegative than A, perturbation arguments give p_B the larger coefficient in $1b_1$ while p_A has the larger weighting in $2b_1$ (5). The related $4a'$ and $5a'$ MOs of pyramidal geometry

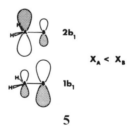

$$2b_1$$
$$X_A < X_B$$
$$1b_1$$

5

might retain similar proportions of p_A and p_B contributions, although possibilities for MO mixing in the lower symmetry nonplanar geometry

TABLE IV

Some calculated structural parameters for H_2AB

	HAH(°)	α(°)	Inversion barrier (kcal/mole)
H_2NO	123	17	0.064
H_2CO^-	116	27	0.875

introduce complications. But assuming that $5a'$ can be satisfactorily represented as in **6** for $X_A < X_B$, we can predict that molecules containing

6

13- or 14-valence electrons should have smaller HAH angles and a greater degree of pyramidalization or out-of-plane folding if the terminal B is more electronegative than the central atom A. Matching ab initio calculations (77) for H_2NO and H_2CO^- conclude that these 13-electron radicals should be nonplanar but with very low barriers to inversion. Table IV contains the calculated inversion barriers, the HAH angles, and the angle α between the A-B bond axis and the bisector of HAH, which measures the degree of out-of-plane folding. With the larger electronegativity difference between central atom A and terminal B, H_2CO^- has the smaller HAH angle, the greater angle of deviation from planarity and the larger inversion barrier as suggested by the properties of $5a'$. The electronegativity difference model predicts an increasing tendency to nonplanarity through the 13-electron series H_2CBr, H_2CCl, H_2CF, and H_2SiF. ESR and infrared spectra show that H_2CF, H_2CCl, and H_2CBr are planar or nearly so. If they were pyramidal with inversion barriers as low as those calculated for H_2NO and H_2CO^-, the nonplanarity would be very difficult to detect experimentally. Ab initio calculations (78) for H_2CF indicate a pyramidal shape with a barrier to inversion of 0.3 to 0.4 kcal/mole. The radical H_2SiF has been shown to be pyramidal on the basis of its ESR spectrum.

REFERENCES

Experimental HAB Structures or Other Data

(1) HCC: W. R. M. Graham, K. I. Dismuke and W. Weltner, Jr., *J. Chem. Phys.* **60**, 3817 (1974); K. D. Tucker, M. L. Kutner, and P. Thaddeus, *Astrophys. J.* **193**, L115 (1974).

(2) HCN$^+$: J. M. Hollas and T. A. Sutherly, *Mol. Phys.* **24**, 1123 (1972).

(3) HNB: E. R. Lory and R. F. Porter, *J. Am. Chem. Soc.* **95**, 1766 (1973).

(4) HNN$^+$: R. D. Smith and J. H. Futrell, *J. Phys. Chem.* **81**, 195 (1977).

(5) HCN: G. Winnewisser, A. G. Maki, and D. R. Johnson, *J. Mol. Spectrosc.* **39**, 149 (1971); G. Herzberg and K. K. Innes, quoted in G. Herzberg. "Molecular Spectra and Molecular Structure," Vol. III, Electronic Spectra and Electronic Structure of Polyatomic Molecules. Van Nostrand-Reinhold, Princeton, New Jersey, 1965.

(6) HNC: D. E. Milligan and M. E. Jacox, *J. Chem. Phys.* **47**, 278 (1967).

(7) HCP: J. W. C. Johns, H. F. Shurvell, and J. K. Tyler, *Can. J. Phys.* **47**, 893 (1969); G. Strey and I. M. Mills, *Mol. Phys.* **26**, 129 (1973).

(8) HBO: E. R. Lory and R. F. Porter, *J. Am. Chem. Soc.* **93**, 6301 (1971).

(9) HBS: E. F. Pearson and R. V. McCormick, *J. Chem. Phys.* **58**, 1619 (1973).

(10) HNN: H. A. Papazian, *J. Chem. Phys.* **32**, 456 (1960).

(11) HCO: J. A. Austin, D. H. Levy, C. A. Gottlieb, and H. E. Radford, *J. Chem. Phys.* **60**, 207 (1974); J. M. Brown and D. A. Ramsay, *Can. J. Phys.* **53**, 2232 (1975).

(12) HCN$^-$: K. D. J. Root, M. C. R. Symons, and B. C. Weatherly, *Mol. Phys.* **11**, 161 (1966); E. L. Cochran, B. C. Weatherly, V. A. Bowers, and F. J. Adrian, *Bull. Am. Phys. Soc.* **13**, 357 (1968).

(13) HBO$^-$: R. C. Catton, M. C. R. Symons, and H. W. Wardale, *J. Chem. Soc. A* 2622 (1969).

(14) HCF: A. J. Merer and D. N. Travis, *Can. J. Phys.* **44**, 1541 (1966).

(15) HCCl: A. J. Merer and D. N. Travis, *Can. J. Phys.* **44**, 525 (1966).

(16) HNO: F. W. Dalby, *Can. J. Phys.* **36**, 1336 (1958); J. L. Bancroft, J. M. Hollas and D. A. Ramsay, *Can. J. Phys.* **40**, 322 (1962).

(17) HPO: M. Lam Thanh and M. Peyron, *J. Chim. Phys.* **61**, 1531 (1964).

(18) HSiCl, HSiBr: G. Herzberg and R. D. Verma, *Can. J. Phys.* **42**, 395 (1964).

(19) HSiI: J. Billingsley, *Can. J. Phys.* **50**, 531 (1972).

(20) HOO: D. E. Milligan and M. E. Jacox, *J. Chem. Phys.* **38**, 2627 (1963); T. T. Paukert and H. S. Johnston, *J. Chem. Phys.* **56**, 2824 (1972); H. E. Hunziker and H. R. Wendt, *J. Chem. Phys.* **60**, 4622 (1974).

(21) HSS: R. K. Gosavi, M. DeSorgo, H. E. Gunning and O. P. Strausz, *Chem. Phys. Lett.* **21**, 318 (1973).

(22) HNF: C. M. Woodman, *J. Mol. Spectrosc.* **33**, 311 (1970).

(23) HOF: H. Kim, E. F. Pearson and E. H. Appelman, *J. Chem. Phys.* **56**, 1, (1972); E. F. Pearson and H. Kim, *J. Chem. Phys.* **57**, 4230 (1972).

(24) HOCl: R. A. Ashby, *J. Mol. Spectrosc.* **23**, 439 (1967); A. M. Mirri, F. Scappini, and G. Cazzoli, *J. Mol. Spectrosc.* **38**, 218 (1971).

(25) HOBr: I. Schwager and A. Arkell, *J. Am. Chem. Soc.* **89**, 6006 (1967).

Calculated HAB Structures

(26) HCC: I. H. Hillier, J. Kendrick, and M. F. Guest, *Mol. Phys.* **30**, 1133 (1975); S. Shih, S. D. Peyerimhoff, and R. J. Buenker, *J. Mol. Spectrosc.* **64**, 167 (1977).

(27) HCN$^+$: S. R. So and W. G. Richards, *J. Chem. Soc. Faraday II* **71**, 62 (1975).

(28) HCN: G. M. Schwenzer, S. V. O'Neil, H. F. Schaefer III, C. P. Baskin, and C. F. Bender, *J. Chem. Phys.* **60**, 2787 (1974); G. M. Schwenzer, C. F. Bender, and H. F. Schaefer, III, *Chem. Phys. Lett.* **36**, 179 (1975); D. Booth and J. N. Murrell, *Mol. Phys.* **24**, 1117 (1972); P. K. Pearson and H. F. Schaefer, III, *J. Chem. Phys.* **62**, 350 (1975).

(29) HNC: D. Booth and J. N. Murrell, *Mol. Phys.* **24**, 1117 (1972); P. K. Pearson and H. F. Schaefer, III, *J. Chem. Phys.* **62**, 350 (1975); G. M. Schwenzer, H. F. Schaefer, and C. F. Bender, *J. Chem. Phys.* **63**, 569 (1975).

(30) HBO: C. Thomson and B. J. Wishart, *Theoret. Chim. Acta* **35**, 267 (1974).

(31) HBS: C. Thomson, *Chem. Phys. Lett.* **25**, 59 (1974); O. Gropen and E. Wisløff-Nilssen, *J. Mol. Struct.* **32**, 21 (1976).
(32) HNN: N. C. Baird, *J. Chem. Phys.* **62**, 300 (1975); K. Vasudevan, S. D. Peyerimhoff, and R. J. Buenker, *J. Mol. Struct.* **29**, 285 (1975).
(33) HCO: P. J. Burna, R. J. Buenker, and S. D. Peyerimhoff, *J. Mol. Struct.* **32**, 217 (1976).
(34) HNO⁺, HON⁺: C. Marian, P. J. Bruna, R. J. Buenker, and S. D. Peyerimhoff, *Mol. Phys.* **33**, 63 (1977).
(35) HBF: D. A. Brotchie and C. Thomson, *Chem. Phys. Lett.* **22**, 338 (1973).
(36) HNO, HON: G. A. Gallup, *Inorg. Chem.* **14**, 563 (1975).
(37) HOO: D. H. Liskow, H. F. Schaefer, III, and C. F. Bender, *J. Am. Chem. Soc.* **93**, 6734 (1971); J. J. Gole and E. F. Hayes, *J. Chem. Phys.* **57**, 360 (1972).
(38) HSS: A. B. Sannigrahi, S. D. Peyerimhoff, and R. J. Buenker, *Chem. Phys. Lett.* **46**, 415 (1977).
(39) HSO: A. B. Sannigrahi, K. H. Thunemann, S. D. Peyerimhoff, and R. J. Buenker, *Chem. Phys.* **20**, 25 (1977).
(40) HOS: A. B. Sannigrahi, S. D. Peyerimhoff, and R. J. Buenker, *Chem. Phys.* **20**, 38 (1977).
(41) HNF: D. A. Brotchie and C. Thomson, *Chem. Phys. Lett.* **22**, 338 (1973).
(42) HOF: R. J. Buenker and S. D. Peyerimhoff, *J. Chem. Phys.* **45**, 3682 (1966).

General References

(43) J. A. Ibers, *J. Chem. Phys.* **40**, 402 (1964); **41**, 25 (1964).
(44) J. C. Evans and G. Y.-S. Lo, *J. Phys. Chem.* **70**, 11 (1966); **71**, 3697 (1967); **73**, 448 (1969).
(45) J. A. Salthouse and T. C. Waddington, *J. Chem. Soc. A* 28 (1966).
(46) P. N. Noble and G. C. Pimentel, *J. Chem. Phys.* **49**, 3165 (1968); V. Bondybey, G. C. Pimentel, and P. N. Noble, *J. Chem. Phys.* **55**, 540 (1971); P. N. Noble, *J. Chem. Phys.* **56**, 2088 (1972).
(47) D. E. Milligan and M. E. Jacox, *J. Chem. Phys.* **53**, 2034 (1970); **55**, 2550 (1971).
(48) L. Andrews, B. S. Ault, J. M. Grzybowski, and R. O. Allen, *J. Chem. Phys.* **62**, 2461 (1975); B. S. Ault and L. Andrews, *J. Chem. Phys.* **63**, 2466 (1975); **64**, 1986 (1976).
(49) P. N. Noble and R. N. Kortzeborn, *J. Chem. Phys.* **52**, 5375 (1970).
(50) C. Thomson, D. T. Clark, T. C. Waddington, and H. D. B. Jenkins, *J. Chem. Soc. Faraday II* **71**, 1942 (1975).
(51) S. A. Harrell and D. H. McDaniel, *J. Am. Chem. Soc.* **86**, 4497 (1964).
(52) G. C. Pimentel and R. D. Spratley, *J. Am. Chem. Soc.* **85**, 826 (1963).
(53) L. C. Allen, R. M. Erdahl, and J. L. Whitten, *J. Am. Chem. Soc.* **87**, 3769 (1965).
(54) P. A. Kollman and L. C. Allen, *J. Chem. Phys.* **51**, 3286 (1969); **52**, 5085 (1970); *J. Am. Chem. Soc.* **92**, 753 (1970).
(55) J. Del Bene and J. A. Pople, *J. Chem. Phys.* **52**, 4858 (1970); B. R. Lentz and H. A. Scheraga, *J. Chem. Phys.* **58**, 5296 (1973).
(56) L. C. Allen, *J. Am. Chem. Soc.* **97**, 6921 (1975); S. Iwata and K. Morokuma, *J. Am. Chem. Soc.* **95**, 7563 (1973).
(57) G. Diercksen and H. Preuss, *Int. J. Quantum Chem.* **1**, 637 (1967).
(58) N. K. Ray, *J. Chem. Phys.* **52**, 463 (1970).
(59) G. Diercksen and H. Preuss, *Int. J. Quantum Chem.* **1**, 641 (1967).
(60) R. Janoschek, G. Diercksen, and H. Preuss, *Int. J. Quantum Chem.* **2**, 159 (1968).

Experimental H_2AB *Structural Data*

(61) H_2CN: E. L. Cochran, F. J. Adrian, and V. A. Bowers, *J. Chem. Phys.* **36**, 1938 (1962); D. Banks and W. Gordy, *Mol. Phys.* **26**, 1555 (1973).

(62) H₂CO⁺: S. P. Mishra and M. C. R. Symons, *Chem. Commun.* 909 (1975).
(63) H₂BO: W. R. M. Graham and W. Weltner, Jr., *J. Chem. Phys.* **65**, 1516 (1976).
(64) H₂CO: K. Takagi and T. Oka, *J. Phys. Soc. Jpn.* **18**, 1174 (1963).
(65) H₂CS: D. R. Johnson, F. X. Powell, and W. H. Kirchhoff, *J. Mol. Spectrosc.* **39**, 136 (1971).
(66) H₂NO: J. Q. Adams, S. K. Nicksic, and J. R. Thomas, *J. Chem. Phys.* **45**, 654 (1966).
(67) H₂CF: R. W. Fessenden and R. H. Shuler, *J. Chem. Phys.* **43**, 2704 (1965); M. E. Jacox and D. E. Milligan, *J. Chem. Phys.* **50**, 3252 (1969).
(68) H₂CCl: M. E. Jacox and D. E. Milligan, *J. Chem. Phys.* **53**, 2688 (1970); L. Andrews and D. W. Smith, *J. Chem. Phys.* **53**, 2956 (1970); J. P. Michaut and J. Roncin, *Chem. Phys. Lett.* **11**, 95 (1971).
(69) H₂CBr: D. W. Smith and L. Andrews, *J. Chem. Phys.* **55**, 5295 (1971).
(70) H₂SiF: M. V. Merritt and R. W. Fessenden, *J. Chem. Phys.* **56**, 2353 (1972).
(71) H₂NCl: D. G. Lister and D. J. Millen, *Chem. Commun.* 1505 (1970).

Calculated H₂AB Structures

(72) H₂CC: A. C. Hopkinson, K. Yates, and I. G. Csizmadia, *J. Chem. Phys.* **55**, 3835 (1971).
(73) H₂NN: N. C. Baird and D. A. Wernette, *Can. J. Chem.* **55**, 350 (1977).
(74) H₂CO: R. Ditchfield, W. J. Hehre, and J. A. Pople, *J. Chem. Phys.* **54**, 724 (1971); B. J. Garrison, H. F. Schaefer, III, and W. A. Lester, Jr., *J. Chem. Phys.* **61**, 3039 (1974).
(75) H₂CF⁺: N. C. Baird and R. K. Datta, *Can. J. Chem.* **49**, 3708 (1971).
(76) H₂BF: M. E. Schwartz and L. C. Allen, *J. Am. Chem. Soc.* **92**, 1466 (1970).
(77) H₂NO, H₂CO⁻: Y. Ellinger, R. Subra, A. Rassat, J. Douady, and G. Berthier, *J. Am. Chem. Soc.* **97**, 476 (1975).
(78) H₂CF: H. Konishi and K. Morokuma, *J. Am. Chem. Soc.* **94**, 5603 (1972).
(79) H₂NF: J. M. Lehn and B. Munsch, *Chem. Commun.* 1062 (1970).
(80) H₂NCl: G. L. Bendazzoli, D. G. Lister, and P. Palmieri, *J. Chem. Soc. Faraday II* **69**, 791 (1973).

6 A_2H_2, A_2H_4, and A_2H_6

The molecular orbitals for the symmetric dimers A_2H_2, A_2H_4, and A_2H_6 can be formed by taking in-phase and out-of-phase combinations of the MOs of their respective monomers AH, AH_2, and AH_3. The orbitals for AH_2 and AH_3 are already available in Chapter 3 and those for AH are very simple. The construction of MOs for large molecules from those of constituent fragments suggests a way to extend qualitative MO theory to larger and larger systems. This chapter deals with molecular shapes, the origin of barriers to internal rotation and inversion, and the dimerization mechanisms of some highly symmetric molecules.

ACETYLENE, DIIMIDE, AND HYDROGEN PEROXIDE

ORBITALS FOR A_2H_2

Figure 1 shows how the MOs for linear HAAH are related to those of 2AH. Only the lowest eight of ten possible MOs appear in Fig. 1. The central A-A bond between two AH monomers resembles that in isoelectronic diatomic molecules. The bonding effect of $1\sigma_g$ is canceled by the antibonding $1\sigma_u$ orbital. In acetylene (ten electrons, isoelectronic with N_2), the three occupied bonding orbitals $2\sigma_g$ and π_u form the triple bond between the two carbons. The ion $HCCH^+$ (1,2) is known to have a longer CC bond (1.26 Å) than that in the neutral molecule (1.21 Å), which is consistent with the removal of an electron from a CC bonding orbital. Diimide N_2H_2 (12

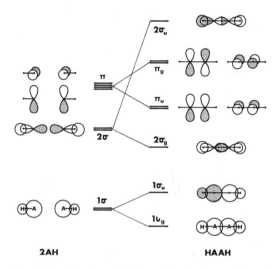

FIG. 1   Formation of MOs of linear A_2H_2 by taking in-phase and out-of-phase combinations of the MOs of two AH fragments.

electrons, isoelectronic with O_2) has two electrons in an antibonding orbital related to π_g and the bond between the two nitrogens is a double bond. Hydrogen peroxide H_2O_2 (14 electrons, isoelectronic with F_2) has an O-O single bond because an electron pair occupies the antibonding MO related to the other π_g orbital. The central bonds lengthen through the series as the A-A bond order decreases: C_2H_2 (1.206 Å) (2), N_2H_2 (1.252 Å) (3), H_2O_2 (1.475 Å) (4). No sixteen-electron dimers with an A-A link, such as HFFH, are possible because $2\sigma_u$ is antibonding. Instead, the $(HF)_2$ dimer has the hydrogen bonded structure HF---HF (5) discussed in Chapter 5. The ESR spectrum of $H_2S_2^-$ indicates that this 15-electron radical ion has nonequivalent hydrogens and nonequivalent sulfurs (6).

Figure 2 displays the valence MOs for linear and nonlinear, cis and trans, geometries of a typical A_2H_2 molecule. Orbital energy changes can be interpreted from overlap changes. A case of noncrossing of orbitals of the same symmetry for cis geometry exactly parallels the HAB model. With 10 electrons, acetylene C_2H_2 is linear. Diimide N_2H_2 (12 electrons) has planar cis and trans isomers (7) because of the lower energies of $2b_2$ (cis) and $3a_g'$ (trans) that stem from the π_{gx} MO of linear geometry. Hydrogen peroxide H_2O_2 (14 electrons) should also be nonlinear. To see why H_2O_2 is actually nonplanar or gauche, look at Fig. 3, which contains pictures of the two highest occupied MOs for H_2O_2 and shows how their energies change as the molecular framework is twisted from planar, trans, through nonplanar, gauche, to planar, cis. For the purposes of Fig. 3, this twisting takes

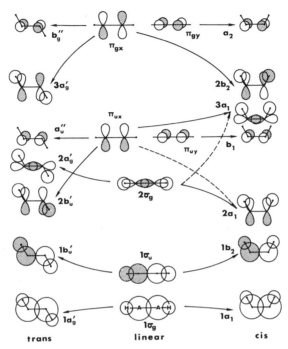

FIG. 2 Qualitative correlation diagram for HAAH in linear and nonlinear (cis and trans) shapes. [Reprinted with permission from Gimarc, *J. Am. Chem. Soc.* **92**, 266 (1970). Copyright by the American Chemical Society.]

place in such a way that the molecular 2-fold symmetry axis remains perpendicular to the page, passing through the midpoint of the A-A bond. Also, the p AOs are locked parallel to the Cartesian axes and are not twisted or rotated. Only the hydrogens move. In the orbital b_g''-b-$2b_2$, the hydrogens move away from the nodal surface in b_g'' (trans) and into positions where their $1s$ orbitals can overlap well with the p orbital lobes in $2b_2$ (cis). This

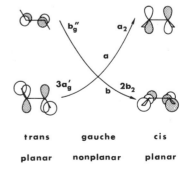

FIG. 3 Correlations for the highest occupied pair of orbitals in H_2O_2 on rotation from trans to cis through gauche or nonplanar geometry. [Reprinted with permission from Gimarc, *J. Am. Chem. Soc.* **92**, 266 (1970). Copyright by the American Chemical Society.]

produces a lowering of energy from b_g'' to b to $2b_2$. At the same time, the hydrogens in $3a_g'$-a-a_2 move out of good overlap in $3a_g'$ (trans) and towards the nodal surface of a_2 (cis) for an energy increase from $3a_g'$ to a to a_2. Now the energy of b_g''-b should drop rapidly because the overlap increases from zero (bottom of cosine curve, Fig. 2, Chapter 3) while the energy of $3a_g'$-a should increase more slowly because the overlap decreases from near maximum (near the top of the cosine curve). Therefore, the energy at which orbitals a and b cross in gauche geometry should be lower than the average of either b_g'' and $3a_g'$ for trans or a_2 and $2b_2$ for cis. While this naive model gives an appealing rationale for the gauche shape of H_2O_2, the actual situation is more complicated. Extensive mixing among orbitals of a symmetry and among orbitals of b symmetry for the gauche shape makes the crossing of these a and b orbitals occur at higher energy. Gauche geometry for H_2O_2 arises from a delicate balance among all the occupied a and b valence orbitals.

EXCITED STATES

Although acetylene is linear in its ground state, its first excited state is nonlinear (trans, C_{2h}) with a considerably longer CC bond. Compare the excited state distance 1.39 Å (8) with 1.21 Å (2) for the ground state. Figure 2 shows that excitation moves an electron from a CC bonding orbital to a CC antibonding orbital producing an increase in bond distance. Furthermore, the orbital (either $3a_g'$ trans or $2b_2$ cis) that receives the excited electron strongly prefers nonlinear geometry. Ab initio calculations predict cis as well as trans excited states for acetylene (9).

Excited trans-N_2H_2 (10) has a considerably larger HNN bond angle (130°) than in the ground state (107°), while the N=N distances in ground and excited states are approximately equal. The excited state is apparently planar although the evidence is not conclusive. In Fig. 3, compare $3a_g'$, the highest occupied ground-state MO, and b_g'', the lowest vacant orbital. Excitation removes an electron from $3a_g'$, the orbital that holds N_2H_2 bent, and adds the electron to b_g'', an orbital with no HNN angle preference. Therefore, one would expect the HNN angle to open on excitation. Since both $3a_g'$ and b_g'' are N-N antibonding, excitation would be expected to have no effect on the N=N distance. Planar shape for the excited state is harder to resolve. Excited N_2H_2 would be similar to H_2O_2, but with only one electron in each a and b orbital in Fig. 3. Therefore, one should expect a gauche or nonplanar structure for excited N_2H_2 but with rotational barriers even smaller than those for H_2O_2. The trans barrier in H_2O_2 is only about 1 kcal/mole (11).

ISOMERIZATION

Two mechanisms seem reasonable for the interconversion of trans and cis isomers of N_2H_2: planar inversion (1) and rotation (2) (through the

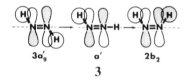

1

2

nonplanar or gauche structure) about $N=N$. In the trans isomer, the higher occupied orbital is $3a_g'$ while in the cis isomer the highest occupied orbital is $2b_2$. See Fig. 3. On rotation from trans to cis through the gauche form, the occupied a MO related to $3a_g'$ (trans) crosses the vacant b MO related to b_g'' (trans). The crossing of occupied and vacant MOs is a violation of the principal of conservation of orbital symmetry and, therefore, a process with a high barrier. Of course, connection of the two isomers through gauche geometry is possible by invoking interaction or superposition of configurations a^2 and b^2, but we can still anticipate a prohibitively large barrier for the rotational process. In the planar inversion mechanism the a, b symmetry is lost and $3a_g'$ can be converted directly into $2b_2$ (3). This process clearly

$$3a_g' \longrightarrow a' \longrightarrow 2b_2$$

3

presents an energy barrier since one hydrogen $1s$ AO is forced to decouple from the lobe of a neighboring p AO and move across a nodal surface of the MO. Still, the symmetry allowed planar inversion process should have a lower barrier than the symmetry forbidden rotational process. Ab initio calculations for N_2H_2 yield a barrier to planar inversion of 47 kcal/mole while even including configuration interaction the rotational barrier is 60 kcal/mole (12). In the excited configuration a^1b^1, the rotational isomerization process should have a very low barrier with the possibility of an energy minimum in the gauche conformation.

ETHYLENE, HYDRAZINE, AND OTHERS

ORBITALS FOR A_2H_4

Qualitative MOs for planar A_2H_4 can be constructed by taking in-phase and out-of-phase combinations of the MOs for separated, bent AH_2 fragments. Figure 4 correlates relative energy levels and orbitals for planar

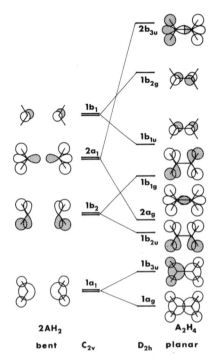

FIG. 4 Formation of MOs of planar A_2H_4 as combinations of MOs of bent AH_2 fragments. [Reprinted from Durig *et al.*, *Vib. Spect. Struct.* **2**, 1 (1973), by courtesy of Marcel Dekker, Inc.]

A_2H_4 and two isolated AH_2 units. The ordering of A_2H_4 energy levels can be rationalized qualitatively. Compared with $2a_1$ (AH_2), the energy of $2a_g$ (A_2H_4) should be low while that of $2b_{3u}$ should be quite high because of the bonding and antibonding p-sigma-type overlaps that result as the two AH_2 fragments are joined. The splittings that occur as a result of the two $1b_2$ and the two $1b_1$ combinations are much smaller because of their pi-type overlap. The A-A antibonding orbital $1b_{1g}$ lies below the bonding orbital $1b_{1u}$. The four $1s$, p overlaps available to $1b_{1g}$ but not to $1b_{1u}$ give $1b_{1g}$ a lower energy. For ethylene (C_2H_4, 12 electrons), the bonding pi-orbital $1b_{1u}$ is the highest occupied MO. The lowest unoccupied MO is the antibonding pi-orbital $1b_{2g}$.

Now consider twisting one AH_2 group relative to the other about the A-A bond from planar D_{2h} to staggered D_{2d} geometry. Figure 5 is the correlation diagram for this process. The orbitals $1a_g$-$1a_1$, $2a_g$-$2a_1$, and $1b_{3u}$-$1b_2$ are axially symmetric and their energies change very little on rotation. Symmetry requires that each b_1, b_2 pair of D_{2h} or D_2 geometry become degenerate in D_{2d}. Both $1b_{1u}$ and $1b_{1g}$ of planar D_{2h} become b_1 in intermediate twisted D_2 geometry. Intended correlations, indicated by dashed lines in Fig. 5, are $1b_{1g}$-$2e$ and $1b_{1u}$-$1e$. Because $1b_{1u}$ lies above $1b_{1g}$, the intended correlations are blocked by the noncrossing rule. It is the rising energy of the MO $1b_{1u}$-b_1-$2e$ that makes ethylene planar (*13*).

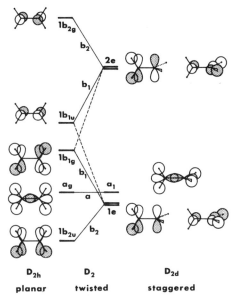

FIG. 5   Qualitative correlation diagram for planar and staggered A_2H_4. The degenerate orbital pair $1e$ has about the same energy as $3a_1$ (staggered). [Reprinted from Durig *et al.*, *Vib. Spect. Struct.* **2**, 1 (1973), by courtesy of Marcel Dekker, Inc.]

From overlap considerations alone one would expect $1e$ to have an energy half-way between $1b_{2u}$ and $1b_{1u}$. However, extended Hückel calculations indicate that this is not the case; instead, $1e$ is lower and $2e$ is higher than the half-way estimates. The mixing or perturbation interaction between orbitals of the same symmetry classification, $1e$ and $2e$, results in a repulsion or divergence of $1e$ and $2e$. Figure 6 shows the details. Although parallel p AOs on the two A atoms must have equal coefficients in b_{2u}, b_{1u}, b_{1g}, and b_{2g} (planar), they need not be equal in $1e$ and $2e$ (staggered). The perturbation interaction or mixing makes each $1e$ orbital mainly a bonding AH_2 MO combined in-phase with a small p AO contribution on the opposite A atom. Each $2e$ orbital becomes primarily a p AO on one A atom interacting out-of-phase with a small AH_2 MO on the other A.

The molecule B_2H_4 (ten valence electrons, not known experimentally) should be staggered rather than planar because electrons in the highest occupied levels $1e$ (D_{2d}) have lower energy than they would in $1b_{2u}$ and $1b_{1g}$ (D_{2h}). The barrier to rotation about the B-B bond should be small. Ab initio SCF MO calculations show that B_2H_4 prefers the staggered D_{2d} conformation *(14)*. The eight-electron molecule Be_2H_4, also unknown, would be flat rather than twisted because $1b_{1g}$ would be empty.

Ground-state ethylene is planar D_{2h} *(13)*. The energy of the highest occupied orbital $1b_{1u}$-b_1-$2e$ increases sharply as the molecule is twisted.

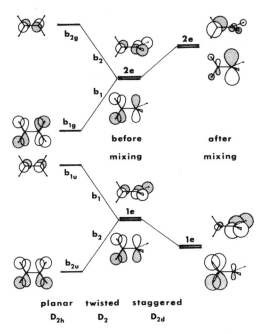

FIG. 6   The mixing of degenerate orbitals $1e$ and $2e$ in staggered D_{2d} geometry. For convenience of representation, orbital b_{1g} is placed above b_{1u}. [Reprinted from Durig *et al.*, *Vib. Spect. Struct.* **2**, 1 (1973), by courtesy of Marcel Dekker, Inc.]

Excited ethylene, singlet or triplet, with one electron occupying $1b_{2g}$-b_2-$2e$ should be staggered D_{2d}, as experiment (*13*) and ab initio calculations (*15*) confirm. Excitation transfers an electron from a C-C bonding to an antibonding orbital that should produce an increase in the C-C bond distance. The ground-state distance is 1.33 Å (*16*). Excited singlet and triplet distances have been calculated as 1.43 and 1.48 Å, respectively (*15*). The ion $C_2H_4^+$ should be planar D_{2h}, a single electron in $1b_{1u}$-b_1-$2e$ being enough to overcome a rather weak net preference of lower orbitals for staggered geometry. Since the electron lost from C_2H_4 to produce $C_2H_4^+$ comes for a C-C bonding orbital, the C-C bond distance in $C_2H_4^+$ should be longer than that in ethylene itself. Calculations give 1.43 Å as the C-C distance in $C_2H_4^+$ (*17*). For the 14-electron molecule N_2H_4, the barrier to rotation about the central N-N bond should be small, but even lower energy shapes are available to this molecule.

Imagine rotating the AH_2 groups of the planar D_{2h} ethylene structure in the plane in a conrotatory fashion to give the planar, hydrogen-bridged D_{2h} shape. Figure 7 is the MO correlation diagram for this process. The A-A distance is assumed constant during the rotation. The energies of the two pi orbitals $1b_{1u}$ and $1b_{2g}$ do not change at all; the hydrogens remain on the

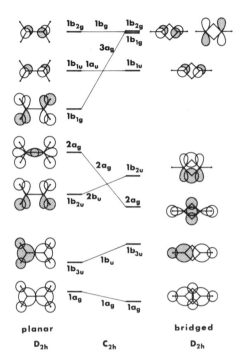

FIG. 7 Qualitative correlation diagram for an A_2H_4 molecule in planar ethylene D_{2h} and hydrogen bridged D_{2h} shapes. [Reprinted from Durig *et al.*, *Vib. Spect. Struct.* **2**, 1 (1973), by courtesy of Marcel Dekker, Inc.]

nodal surface during the rotation. The energy of $1b_{2u}$ changes only slightly despite the rather large changes in overlaps among constituent AOs. Of the four hydrogen $1s$ AOs in the ethylenic form of $1b_{2u}$, two disappear on rotation to terminal positions on the nodal surface in the bridged structure. The other two $1s$ AOs move into bridging positions where each overlaps both p AOs. In the ethylene form there are four $1s$, p overlaps of about 87% of the maximum and in the bridged form there are four $1s$, p overlaps of about 70%, the overlap decrease producing a slightly higher energy for $1b_{2u}$ in the bridged structure. Only two orbitals, $2a_g$ and $1b_{1g}$, change much in energy. The $2a_g$ orbital shows a sizeable energy decrease on going to the bridged shape. Of four $1s$, p overlaps, each at about 50% of maximum in the ethylenic shape, two increase to 100% (at the terminals) and two increase to 70% (bridging) with both p AOs. It is the $2a_g$ orbital that gives the eight-electron Be_2H_4 molecule the bridged D_{2h} shape. Indeed, ab initio SCF MO calculations for Be_2H_4 indicate a planar bridged structure (*18*).

On rotation from ethylenic to bridged shapes the $1b_{1g}$ MO rises sharply and becomes degenerate with $1b_{2g}$, with all four hydrogens moving onto

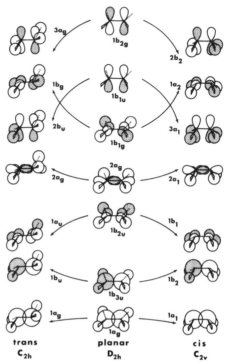

FIG. 8 Qualitative orbital correlations for an A_2H_4 molecule in planar and nonplanar (cis and trans) shapes. [Reprinted from Durig *et al.*, *Vib. Spect. Struct.* **2**, 1 (1978), by courtesy of Marcel Dekker, Inc.]

nodal surfaces. It is $1b_{1g}$ that prevents molecules with more than eight valence electrons, such as B_2H_4, C_2H_4, and N_2H_4, from assuming the bridged D_{2h} shape.

Next, consider bending or folding the AH_2 groups out of the plane of the ethylene structure either towards each other (cis, C_{2v}) or away from each other (trans, C_{2h}). Figure 8 shows the orbital energy changes for these processes. If $1b_{1u}$ and $1b_{1g}$ of planar D_{2h} geometry are close in energy as they appear to be in Fig. 4, then $1b_{1g}$ and $2b_u$ (trans) will surely cross when the structure is folded. The same should be true of $1a_2$ and $3a_1$ (cis). The orbitals $1b_g$ (trans) and $1a_2$ (cis) rise in energy as the terminal hydrogens move towards the nodal surface that cuts between the two AH_2 groups. In ethylene, the highest occupied orbital is $1b_{1u}$. Of the six occupied MOs, three increase in energy due to increased out-of-phase overlaps while three decrease in energy because of in-phase overlaps. Since an out-of-phase interaction raises the energy more than a comparable in-phase interaction lowers it, out-of-plane folding raises the total energy for ethylene but the net energy stabilization of the planar shape is really rather small. In fact, the CH_2 rock, the out-of-plane folding vibration, is the lowest energy vibration in the ethylene spectrum. $C_2H_4^+$ should be stiffer than ethylene to this folding motion or rock because the lost electron comes from an orbital that favors folding.

Hydrazine N_2H_4, with an electron pair in $3a_g$-$1b_{2g}$-$2b_2$ should be non-planar cis or trans rather than planar. Weaker end-end out-of-phase interactions in $3a_g$ (trans) compared to $2b_2$ (cis) give the trans conformation a slightly lower energy than the cis. As will be discussed shortly, hydrazine actually has twisted or gauche geometry (19).

Should excited ethylene be staggered D_{2d} or nonplanar trans? Compare Figs. 5 and 8. For the planar-staggered scheme (Fig. 5), excitation takes an electron from an orbital that holds the molecule planar ($1b_{1u}$-$2e$) and adds it to one favoring staggered geometry ($1b_{2g}$-$2e$). In the planar-trans scheme (Fig. 8), excitation moves an electron between orbital systems that both favor folding, although $3a_g$-$1b_{2g}$-$2b_2$ probably favors folding a bit more than $2b_u$-$1b_{1u}$-$3a_1$. The larger change favors the staggered form for excited ethylene.

Figure 9 shows how molecular orbitals and their energies change when the molecular framework is twisted from C_{2h} trans through the C_2 gauche form to C_{2v} cis. Figure 9 can be constructed from the MO pictures for trans and cis in Fig. 8 by connecting the trans orbitals of a symmetry with the

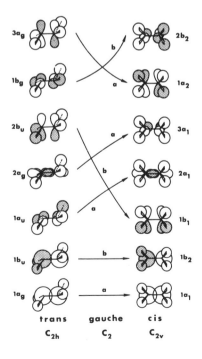

FIG. 9 Qualitative correlations relating A_2H_4 MOs of the nonplanar trans conformation with those of the cis form by rotation through nonplanar gauche geometry. In both sets of MO pictures, the C_2 symmetry axis is perpendicular to the page and passes through the midpoint of the A–A bond. [Reprinted from Durig *et al.*, *Vib. Spect. Struct.* **2**, 1 (1978), by courtesy of Marcel Dekker, Inc.]

cis orbitals of a symmetry and the trans b MOs with the cis b MOs such that a does not cross a nor b cross b. (Crossings of a and b are, of course, allowed.) For the pictures of MOs of both cis and trans geometries in Fig. 9, the C_2 rotational symmetry axis is perpendicular to the plane of the page and passes through the midpoint of the A-A bond. Imagine carrying out the twisting without moving the C_2 axis and without tilting or twisting the p AOs in the A atoms. The MOs $2a_g$ (trans) and $2a_1$ (cis) should be close in energy because of their similar AO compositions. Both are classified as a symmetry in intermediate gauche geometry. However, orbital symmetry prevents them from being connected in Fig. 9. Instead, $2a_1$ is linked to $1a_u$ below $2a_g$. The MOs containing p orbitals perpendicular to the A-A axis (pi-type) undergo sizable energy changes. To interpret these changes it is necessary to appreciate the three different kinds of overlap that occur between A atom p and hydrogen $1s$ AOs. These overlaps are shown in Fig. 10 along with the percentages of maximum overlap, assuming an HAH angle of 110° and an out-of-plane folding angle of 60°. The trans $3a_g$ MO with 41 % overlap per hydrogen becomes $1a_2$ in cis geometry for which each hydrogen has 82 % overlap. This overlap increase causes the energy of $3a_g$-$1a_2$ to fall for trans to cis rotation. Similarly, $1b_g$ (trans) with 82 % overlaps rotates to $2b_2$ (cis) with 41 % overlaps and a higher energy. Thus, the crossing of $3a_g$-$1a_2$ and $1b_g$-$2b_2$ is quite predictable. Recall that the cosine function changes faster for angles near 90° (small p, $1s$ overlaps) than for angles near 0° (large overlaps). If no other conditions intervened, the energy of $3a_g$-a would drop faster than that of $1b_g$-b would rise below it, giving a lower total energy for gauche geometry. A similar crossing of highest occupied a and b orbitals occured in H_2O_2 providing a similar rationalization for the gauche geometry of that molecule. One should note, however, that the situation is more complicated than is represented in Fig. 9. In the lower C_2 symmetry, extensive mixing occurs among orbitals of a classification and among those of the b classification, raising the energies of the higher energy MOs and lowering those below. It is the delicate balance that arises from all of these orbitals that gives N_2H_4 its minimum energy in the gauche C_2 conformation.

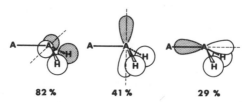

82 %   41 %   29 %

FIG. 10   Percentages of maximum overlap between hydrogen $1s$ and adjacent A atom p orbitals. [Reprinted from Durig *et al.*, *Vib. Spect. Struct.* **2**, 1 (1978), by courtesy of Marcel Dekker, Inc.]

DIMERIZATION PROCESSES

Reactions between two AH_2 groups to form an A_2H_4 molecule are known to occur. Although it might seem reasonable to suppose that the geometrical arrangements or approaches of a pair of AH_2 monomers are simply the ones that lead directly to the lowest energy geometry of the A_2H_4 product, all such least-motion approaches may not occur with low activation energies.

Figure 4 is a correlation diagram for the coplanar approach of two AH_2 fragments that might describe the dimerization of CH_2 to form planar ethylene. For triplet ground-state methylene, with electron configuration $(1a_1)^2(1b_2)^2(2a_1)^1(1b_1)^1$, the electrons flow smoothly and directly from the occupied MOs of the reactants into the lowest available MOs of ethylene. This reaction should be highly exothermic since electrons from $2CH_2$ experience a net energy lowering as they form the sigma and pi bonds of C_2H_4. The activation energy should be zero or negligible. For singlet methylene the electron configuration is $(1a_1)^2(1b_2)^2(2a_1)^2$. A coplanar, least-motion approach of two singlet methylenes would require a large activation energy because it would lead to ethylene with the $1b_{1u}$ and $1b_{2g}$ orbitals empty and the antibonding $2b_{3u}$ orbital doubly occupied. Ab initio SCF MO calculations including configuration interaction confirm that the coplanar dimerization of two singlet methylenes has a significant activation barrier (20). Therefore, singlet methylene must dimerize by some other process. The methylene dimerization has been studied by extended Hückel calculations, the results of which show that the singlet methylene dimerization should occur through the nonleast-motion approach pictured in Fig. 11 (21). This process is an attack by the lone pair of electrons in $2a_1$ of one methylene on the empty pi-orbital $1b_1$ of the other CH_2 (4). That this must be so can be seen from orbital pictures and simple symmetry arguments. The coplanar approach for the dimerization of singlet methylenes fails

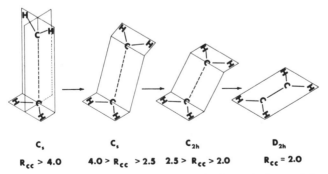

C_s C_s C_{2h} D_{2h}

$R_{cc} > 4.0$ $4.0 > R_{cc} > 2.5$ $2.5 > R_{cc} > 2.0$ $R_{cc} = 2.0$

FIG. 11 Hoffmann's nonleast-motion C_s approach of two singlet methylenes to form ground-state ethylene, as revealed by extended Hückel calculations. (From Ref. 48.)

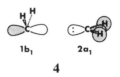

4

because it offers no direct or easy way to feed electrons into the ethylene $1b_{1u}$ MO, which is the in-phase combination of two empty $1b_1$ orbitals of the CH_2 fragments. To fill $1b_{1u}$, the CH_2 reactants must approach each other in a way such that the empty $1b_1$ MOs of each CH_2 can mix with the occupied orbitals on the other CH_2. This cannot happen in the coplanar approach because the empty $1b_1$ orbitals are antisymmetric with respect to the molecular plane while the occupied orbitals $1a_1$, $1b_2$, and $2a_1$ are symmetric. In other words, $1b_1$ is orthogonal to the occupied orbitals of the other methylene. Two possibilities come to mind that allow $1b_1$ to mix. First is a staggered D_{2d} approach followed by a rotation about the C-C axis to give planar D_{2h} ethylene. In the staggered approach the empty $1b_1$ on each methylene can mix with the filled $1b_2$ on the other (**5**). However, in-phase

$$1b_2 \quad + \quad 1b_1 \quad \longrightarrow \quad$$
$$1b_1 \quad + \quad 1b_2 \quad \longrightarrow \quad$$

$$\left. \begin{array}{c} \\ \\ \end{array} \right\} 1e(D_{2d}) \longrightarrow 1b_{1g} \text{ and } 1b_{2u}$$
$$(D_{2h})$$

5

and out-of-phase combinations of the two filled $2a_1$ orbitals of the monomers produce filled $2a_1$ and $2b_2$ MOs of the dimer (**6**). The $2b_2$ (C_2H_4, D_{2d})

$$\qquad \pm \qquad \longrightarrow 2a_1 \text{ and } 2b_2 \ (D_{2d})$$

6

orbital is antibonding and on rotation to planar geometry it converts to the antibonding $2b_{3u}$ (D_{2h}) orbital, still leaving $1b_{1u}$ (D_{2h}) empty and giving the product the same high-energy electron configuration as the coplanar approach. Thus, orbital symmetry rules out the staggered D_{2d} approach for singlet methylene dimerization.

Another approach through which empty $1b_1$ MOs on each CH_2 could mix with lower energy occupied orbitals on its reacting partner is the perpendicular C_s approach shown in Fig. 11. In this orientation $1b_1$ of each fragment mixes with $2a_1$ of the other. From C_s geometry through nonplanar C_{2h} to planar D_{2h}, electrons flow smoothly to the lowest available orbitals of planar ethylene.

Although the staggered D_{2d} approach is not allowed for singlet methylenes, it is allowed for the dimerization of two BH_2 radicals. In this case, the in-phase combination of singly occupied $2a_1$ (BH_2) orbitals leads to a filled and bonding $2a_1$ (B_2H_4) orbital of the dimer. The antibonding $2b_2$ (B_2H_4) is empty and the least-motion process is allowed.

The dimerization of two linear BeH_2 monomers can occur through a planar C_{2h} least-motion process that leads directly to the bridged D_{2h} structure for the Be_2H_4 dimer. Figure 12 is the appropriate correlation diagram. For a linear AH_2 molecule (four electrons), the $2a_1$ and $1b_1$ (C_{2v}) orbitals constitute the degenerate pair $1\pi_u(D_{\infty h})$. Dashed tie-lines in the planar C_{2h} region of Fig. 12 indicate the considerable mixing of $2\sigma_u$ and the components of $1\pi_u$ that lie in the molecular plane. Solid lines show actual orbital connections that are channels for electron flow from reactants to product. Eight valence electrons in two BeH_2 molecules feed directly into the lowest available energy levels of bridged Be_2H_4 and the least-motion approach is allowed. In Fig. 12 the two BeH_2 units are coplanar and parallel but tilted with respect to the axis connecting the two Be atoms. For con-

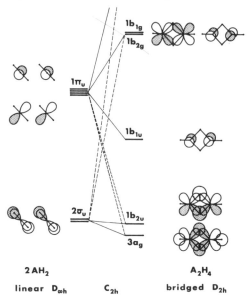

FIG. 12   Extended Hückel MO correlations for the least-motion approach of two AH_2 fragments to form a bridged A_2H_4 dimer. Solid lines show actual connections between orbital energies of reactants and products; dashed lines indicate orbital mixing. [Reprinted with permission from Gimarc, *J. Am. Chem. Soc.* **95**, 1417 (1973). Copyright by the American Chemical Society.]

venience in constructing the MO pictures for Be_2H_4 from the orbitals of the BeH_2 fragments in such a relationship, the Be_2H_4 orbitals are represented in an alternative but equivalent AO basis set in which the axes of two of the Be atom p AOs are tilted by 45° with respect to those in Fig. 7. The $2\sigma_u$ and $1\pi_u$ orbitals on the two BeH_2 fragments mix in bridged geometry to form $2a_g$, which is responsible for the net stabilization of Be_2H_4 with respect to $2BeH_2$. Since the two $2\sigma_u$ components in $2a_g$ do not point directly at each other and the $1\pi_u$ components mix in from rather high energy, the energy of $2a_g$ (Be_2H_4) is not a great deal lower than that of $2\sigma_u$ (BeH_2). Consequently, the bridge bond of Be_2H_4 should be rather weak. Ab initio SCF MO calculations give a dissociation energy of around 31 kcal/mole for the Be_2H_4 bridge bond (18).

NH$_2$ radicals dimerize to form hydrazine, although other products are known to result and, depending on conditions, may be more important (22). Ground-state NH_2 radicals have the electron configuration $(1a_1)^2(1b_2)^2(2a_1)^2(1b_1)^1$. A study of Fig. 4 reveals that the coplanar D_{2h} approach is not allowed for ground-state dimerization of NH_2. The least-motion approach leading directly to gauche hydrazine is allowed (23). For representational convenience we will consider a related allowed process, the trans C_{2h} approach in which two NH_2 radicals come together in parallel planes. The in-phase combination of monomer half-filled $1b_1$ MOs leads to the filled and bonding $2a_g$ (trans) sigma-type MO of the dimer (7). In-phase and out-of-phase combinations of the doubly occupied $2a_1$ MOs of

7

monomer lead to the bonding and antibonding pi-type MOs $2b_u$ and $3a_g$ of N_2H_4 (**8**). The process is allowed. Similar arguments hold for the C_2 or gauche approach, which should have slightly lower energy.

8

ETHANE AND DIBORANE

ORBITALS FOR A_2H_6

Figure 13 shows the formation of valence MOs for staggered (D_{3d}) A_2H_6 from in-phase and out-of-phase combinations of the orbitals of two planar

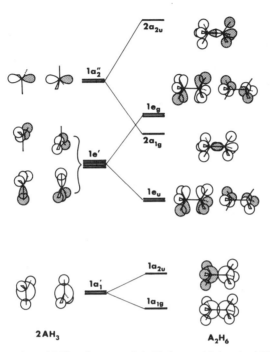

FIG. 13   Formation of MOs of staggered A_2H_6 by combining the MOs of two planar AH_3 fragments. [Reprinted with permission from Gimarc, *J. Am. Chem. Soc.* **95**, 1417 (1973). Copyright by the American Chemical Society.]

AH_3 (D_{3h}) fragments. The relative order of the A_2H_6 energy levels in Fig. 13 generally follows in a straightforward way from qualitative considerations alone. Only in the case of the relative positions of the $1e_g$ and $2a_{1g}$ orbitals is there any uncertainty about energy order (*24*). These two orbitals should be close in energy and their relative position should be highly dependent on the A-A distance; the closer the two principal atoms, the lower the energy of the in-phase combination $2a_{1g}$ and the higher the energy of the out-of-phase combination $1e_g$.

Figure 14 correlates orbitals and energies of A_2H_6 between the staggered ethane structure (D_{3d}) and the bridged diborane shape (D_{2h}) (**9**). Only two orbitals undergo significant energy changes: $2a_{1g}$-$2a_g$ and $1e_g$-$1b_{2g}$. On rocking from the staggered form to the bridged shape, four hydrogens in

9

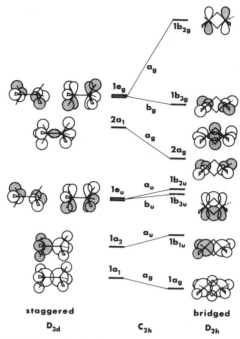

staggered   bridged

D_{3d} C_{2h} D_{2h}

FIG. 14   Qualitative MO correlations for A_2H_6 in staggered D_{3d} and bridged D_{2h} shapes. [Reprinted with permission from Gimarc, *J. Am. Chem. Soc.* **95**, 1417 (1973). Copyright by the American Chemical Society.]

$2a_{1g}$ move to terminal positions and two to bridging positions. The $1s$ AOs on the four terminal hydrogens in the D_{2h} structure are in more favorable overlap with the p-sigma AOs on the principal atoms than they were in the D_{3d} form (from 32% of maximum in D_{3d} to nearly 50% in D_{2h}). Figure 15 shows overlap percentages between p orbitals and hydrogen $1s$ orbitals disposed at tetrahedral angles about the A atom. The $1s$ AOs of the two

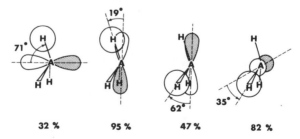

32 %   95 %   47 %   82 %

FIG. 15   Percentages of maximum overlap and angles between p orbital axes and A–H bonds for hydrogens tetrahedrally disposed about an atom A. [Reprinted with permission from Gimarc, *J. Am. Chem. Soc.* **95**, 1417 (1973). Copyright by the American Chemical Society.]

hydrogens that move to the bridge start from 32% of maximum in $2a_{1g}$ and move to locations in $2a_g$ offering each about 70% of maximum overlap with two p-sigma orbitals. This amounts to a considerable increase in in-phase overlap among constituent atomic orbitals in $2a_g (D_{2h})$ relative to $2a_{1g} (D_{3d})$, lowering the energy of $2a_g$ as Fig. 14 shows. Next, consider the $1b_{2g} (D_{2h})$ orbital of the bridged shape. In its $1e_g (D_{3d})$ form in staggered geometry, two hydrogen $1s$ orbitals overlap at about 95% of maximum with p-pi AOs while each of four others overlap about 50%. In the bridged structure all six hydrogens lie on nodal surfaces of $1b_{2g}$, eliminating all hydrogen $1s$ contributions to $1b_{2g}$ and greatly raising the energy of $1b_{2g}$ relative to $1e_g$. The destabilization of $1b_{2g}$ is considerably greater than the stabilization of $2a_g$. Energy and overlap changes of other MO systems are either small or they mutually cancel each other. For example, four hydrogen $1s$ orbitals in $1e_u$ (D_{3d}) $(47\%$ overlap) drop out of $1b_{3u} (D_{2h})$ all together; the terminal hydrogens lie on a nodal surface. Meanwhile, two of the $1s$ orbitals in $1e_u$, already in good overlap, move to comparable positions in $1b_{3u}$, but each $1s$ orbital overlapping two p-pi orbitals in $1b_{3u}$. The result is nearly constant energy for $1e_u$-$1b_{3u}$.

Only two MOs in Fig. 13 affect the structures of ethane and diborane. Diborane, with 12 valence electrons, has $1b_{3g}$ as its highest occupied MO. The $2a_{1g}$-$2a_g$ orbital is, therefore, responsible for holding B_2H_6 in the bridged shape (25). The highest occupied orbital in ethane is $1e_g$-$1b_{2g}$, the steeply rising energy of which forces ethane to be staggered (26) rather than bridged. The 13-electron radical ions $B_2H_6^-$ and $C_2H_6^+$ are known experimentally. The $1e_g$-$1b_{2g}$ orbital rises so steeply that if it contains even one electron it should suffice to hold these ions in the staggered form. Both ab initio calculations and experiment indicate ethanelike rather than bridged shapes for these ions (27–30).

These arguments have neglected two differences between the structures of B_2H_6 and C_2H_6. First, the distance between the principal atoms was assumed to remain constant. In fact, the B-B bond distance in diborane (25) is 0.23 Å longer than the C-C bond length in ethane (26). Lengthening the A-A bond distance in the bridged structure would have the effect of lowering the energies of all the A-A antibonding orbitals while raising those of the bonding orbitals. The net energy result would be small compared to that produced by the angular variations and the molecular shape conclusions would remain unchanged. The other neglected difference is opening of the angle between the terminal hydrogens H_t from $\angle\ HCH = 109°$ in ethane to $\angle\ H_t BH_t = 121°$ in diborane. The 12° angle change produces overlap changes amounting to 5 to 8%, negligible compared to other overlap differences between the D_{3d} and D_{2h} structures and too small to affect the major conclusions about molecular shape.

METHYL RADICAL AND BORANE DIMERIZATION PROCESSES

Both methyl radical and borane dimerize readily through exothermic reactions with zero or negligible activation energies (*31,32*). The energy change for $2CH_3 \rightarrow C_2H_6$ is the C-C bond energy, D (H_3C-CH_3), which is in the range 85 to 90 kcal/mole (*32*). The symmetrical diborane bridge bond energy D (H_3B-BH_3) has been estimated indirectly in various ways, yielding results that differ widely, but likely assumptions lead to values around 22 and 35 kcal/mole (*33–37*). An ab initio SCF MO calculation plus an estimate of the error due to electron correlation gives 36 kcal/mole (*36*). Despite a rather large uncertainty, all estimates agree that D (H_3B-BH_3) is considerably less than D (H_3C-CH_3).

Consider two possible transition states through which two planar AH_3 units might approach each other (**10**). The D_{3d} process is the least-motion

10

approach of two methyl radicals to form staggered ethane described in Fig. 13. Seven electrons in each of two methyl radicals can flow smoothly and directly from reactant energy levels to the lowest available energy levels of ethane. Figure 13 shows that the total energy should steadily decrease (zero activation energy, exothermic) as electrons in singly occupied $1a_2''$ orbitals of $2CH_3$ fall in energy to become an electron pair in $2a_{1g}$ of ethane. It is conceivable that two borane molecules might approach each other in the staggered orientation and then rearrange to form the bridged structure. However, the D_{3d} approach is not favorable for two BH_3 molecules with six electrons each. An electron configuration $(1a_1')^2(1a_1')^2(1e')^4(1e')^4$ for $2BH_3$ would lead to $(1a_{1g})^2(1a_{2u})^2(1e_u)^4(1e_g)^4$ for staggered B_2H_6 with $2a_{1g}$ empty. For normal or greater B-B separations, $2a_{1g}$ may be above $1e_g$ anyway, so a violation of orbital symmetry conservation might not occur. Still, this process would be uphill in energy all the way as the two boranes came together. For each occupied B-B bonding orbital there would be a corresponding occupied antibonding orbital of similar composition and the energy lowering due to the bonding orbitals could be expected to be more than offset by the energy increase resulting from the antibonding MOs. Hence, the D_{3d} transition state can be ruled out for the borane dimerization.

The C_{2h} process represents the least-motion approach of two planar BH_3 molecules to form bridged B_2H_6 directly. Figure 16 is the appropriate correlation diagram. The alternative AO basis set has been used in Fig. 16 for

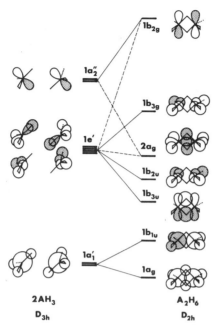

FIG. 16 Qualitative MO correlations for the association of two planar AH_3 molecules through the C_{2h} approach. [Reprinted with permission from Gimarc, *J. Am. Chem. Soc.* **95**, 1417 (1973). Copyright by the American Chemical Society.]

convenience in representing the MO pictures. Examples of A_2H_6 (D_{2h}) MOs in this basis set appear in **11**. Dashed lines in Fig. 16 indicate the considerable

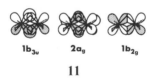

11

mixing of $1e'$ and $1a_2''$ orbitals from isolated AH_3 fragments to make $1b_{3u}$, $2a_g$, and $1b_{2g}$ MOs for A_2H_6 in D_{2h} geometry. This mixing allows six electrons from each of two BH_3 molecules to flow smoothly into the lowest available energy levels for bridged B_2H_6. The total energy should decrease steadily as the two BH_3 units approach; the activation energy should be zero. Ab initio SCF MO calculations find that the C_{2h} approach is much preferred over an unsymmetrical C_s transition state involving a single bridging hydrogen (*38*). The energy of $2a_g$ (B_2H_6) is comparable to that of $1e'$ (BH_3). The stabilization of B_2H_6 provided by $1b_{2u}$ is at least canceled by the antibonding form $1b_{3g}$. Therefore, the net energy lowering, D (H_3B-BH_3), comes mainly from the formation of $1b_{3u}$. This is a pi-type MO and it should provide less stabilization for B_2H_6 compared to $2BH_3$ than would the sigma-type $2a_{1g}$ for C_2H_6 relative to $2CH_3$, or D (H_3B-BH_3) < D (H_3C-CH_3), which is known to be true experimentally.

There is experimental evidence to suggest that D (H_3B^+-BH_3) > D (H_3B-BH_3) (39). This is in accord with the qualitative MO model (Fig. 16); the electron lost to make $B_2H_6^+$ comes from the B-B antibonding MO $1b_{3g}$, which is clearly higher in energy than the highest occupied MOs ($1e'$) of the BH_3 fragments.

BARRIERS TO INTERNAL ROTATION

Figure 17 displays the occupied MOs of ethane in eclipsed (D_{3h}) and staggered (D_{3d}) conformations. Orbital energy changes are quite small compared to previous examples because overlap changes are those between hydrogen AOs at opposite ends of the molecule. The three nondegenerate orbitals with full 3-fold symmetry ($1a_1'$-$1a_{1g}$, $1a_2''$-$1a_{2u}$, and $2a_1'$-$2a_{1g}$) experience almost no energy changes on going from eclipsed to staggered geometry. The largest energy changes appear in $1e'$-$1e_u$ and $1e''$-$1e_g$, which are, respectively, C-C bonding and antibonding pi-type MOs. Since anti-

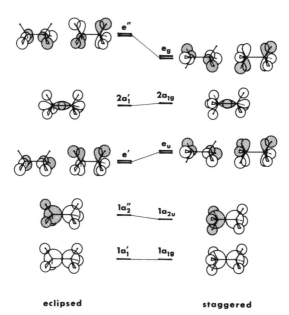

eclipsed   **staggered**

FIG. 17   Orbitals for eclipsed and staggered ethane. Energy changes are quite small for MOs that are symmetric with respect to the 3-fold C-C axis. [Reprinted with permission from Gimarc, *J. Am. Chem. Soc.* **95**. 1417 (1973). Copyright by the American Chemical Society.]

bonding interactions always produce energy increases that are larger than the decreases caused by the corresponding in-phase interactions, the anti-bonding $1e''$-$1e_g$ orbitals determine the staggered conformation of ethane, reducing the out-of-phase or repulsive interactions (40). If the rotational conformation of ethane follows the phase relationships of the highest occupied MOs, conformations of other molecules may follow a similar rule. We have already seen that this is the case for hydrogen peroxide and hy-drazine. In methylamine H_3CNH_2 the orbitals a' and a'' are related to the $1e''$-$1e_g$ MOs of ethane (12). Of course, a' and a'' are not degenerate, but

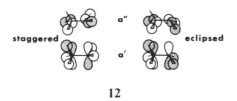

12

these orbitals give methylamine a staggered conformation for the same reasons that $1e''$-$1e_g$ make ethane staggered. Notice that a'' has the same AO composition as its relative in $1e''$-$1e_g$, but a' lacks a hydrogen $1s$ AO. The missing AO in a' reduces out-of-phase overlaps more in eclipsed geometry (where they were greater to begin with) than they do in the staggered con-formation, and, therefore, the barrier to rotation in methylamine should be less than that in ethane. The ethane barrier is 2.93 kcal/mole (41) compared to 1.98 kcal/mole in methylamine (42). For methyl alcohol H_3COH the comparable a' and a'' MOs are shown in **13**. Although a'' no longer shows a

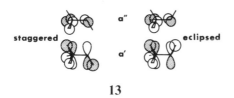

13

conformational preference, the a' MO has lower out-of-phase overlaps in staggered geometry. Methyl acohol should be staggered but with a rotational barrier lower than that of methylamine because only a' acts to determine the conformation. The methyl alcohol barrier is 1.07 kcal/mole (43).

The a''-type MO of ethane has been compared to the higher occupied pi-orbital ϕ_2 of butadiene $H_2C{=}CH{-}CH{=}CH_2$, with the $1s$ AOs of the methyl hydrogens in ethane acting as a hyperconjugative extension of the pi-orbital of butadiene (*40,44*). Out-of-phase overlaps between terminal *p*s in ϕ_2 give butadiene a trans conformation (**14**).

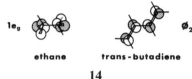

1e_g ϕ_2

ethane   trans-butadiene

14

The highest occupied MO of propene H₃C—CH=CH₂ (**15**) is also

15

similar to the ethane $1e_g(a'')$ orbital. (In constructing the highest occupied a'' orbital, remember that a C=C double bond requires that the two carbon atom *p* AOs have the same phase, but for a C-C single bond the carbon *p* AOs must be of opposite phase.) Again, out-of-phase interactions between methyl hydrogens and the *p* AO on the terminal carbon of the double bond are minimized in the conformation shown above, with a methyl hydrogen eclipsing the double bond.

Not all conformations are determined by repulsions or out-of-phase overlaps. In the highest occupied MO of methyl formate H₃COCHO (**16**),

16

the methyl hydrogen AOs have the same phase as the *p* AO on the carbonyl oxygen. The closest approach of the two methyl hydrogens to the *p* AO of the carbonyl oxygen gives maximum overlap and a conformation in which the methyl hydrogens are staggered with respect to the carbonyl group. In dimethyl acetylene H₃C—C≡C—CH₃, 3-fold symmetry again gives doubly degenerate highest occupied MOs, shown in **17** in eclipsed geometry. End-end in-phase overlaps of the methyl hydrogens prefer the eclipsed rather than the staggered conformation here.

The properties of the highest occupied MOs can rationalize trends in rotational barriers through numerous series, such as the decreasing trend

17

through isobutene, *trans*-2-butene, and *cis*-2-butene (C_4H_8). The qualitative model has been refined to account for trends in barrier heights for methyl groups attached to polar bonds (*45*).

REFERENCES

(1) J. M. Hollas and T. A. Sutherley, *Mol. Phys.* **21**, 183 (1971).

(2) M. T. Christensen, D. R. Eaton, B. A. Green, and H. W. Thompson, *Proc. Roy. Soc. London* **A238**, 15 (1956); W. J. Lafferty and R. J. Thibault, *J. Mol. Spectrosc.* **14**, 79 (1964).

(3) M. Carlotti, J. W. C. Johns, and A. Trombetti, *Can. J. Phys.* **52**, 340 (1974).

(4) W. C. Oelfke and W. Gordy, *J. Chem. Phys.* **51**, 5336 (1969); W. R. Busing and H. A. Levy, *J. Chem. Phys.* **42**, 3054 (1965); E. Prince, S. F. Trevino, C. S. Choi, and M. K. Farr, *J. Chem. Phys.* **63**, 2620 (1975).

(5) T. R. Dyke, B. J. Howard, and W. Klemperer, *J. Chem. Phys.* **56**, 2442 (1972).

(6) M. J. Lin and J. H. Lunsford, *J. Phys. Chem.* **80**, 2015 (1976).

(7) K. Rosengren and G. C. Pimentel, *J. Chem. Phys.* **42**, 507 (1965); V. E. Bondybey and J. W. Nibler, *J. Chem. Phys.* **58**, 2125 (1973).

(8) K. K. Innes, *J. Chem. Phys.* **22**, 863 (1954).

(9) D. Demoulin, *Chem. Phys.* **11**, 329 (1975).

(10) R. A. Back, C. Willis, and D. A. Ramsay, *Can. J. Chem.* **52**, 1006 (1974).

(11) R. H. Hunt, R. A. Leacock, C. W. Peters, and K. T. Hecht, *J. Chem. Phys.* **42**, 1931 (1965).

(12) N. W. Winter and R. M. Pitzer, *J. Chem. Phys.* **62**, 1269 (1975).

(13) R. S. Mulliken, *Rev. Mod. Phys.* **14**, 265 (1942); A. J. Merer and R. S. Mulliken, *Chem. Rev.* **69**, 639 (1969); R. S. Mulliken, *J. Chem. Phys.* **66**, 2448 (1977).

(14) J. D. Dill, P. v. R. Schleyer and J. A. Pople, *J. Am. Chem. Soc.* **97**, 3402 (1975).

(15) R. J. Buenker and S. D. Peyerimhoff, *Chem. Phys.* **9**, 75 (1975).

(16) K. Kuchitsu, *J. Chem. Phys.* **44**, 906 (1966).

(17) W. A. Lathan, W. J. Hehre, and J. A. Pople, *J. Am. Chem. Soc.* **93**, 808 (1971).

(18) R. Ahlrichs, *Theoret. Chim. Acta* **17**, 348 (1970).

(19) Y. Morino, T. Iijima, and Y. Murata, *Bull. Chem. Soc Jpn.* **33**, 46 (1960); J. R. Durig, S. F. Bush, and E. E. Mercer, *J. Chem. Phys.* **44**, 4238 (1966); F. G. Baglin, S. F. Bush, and J. R. Durig, *J. Chem. Phys.* **47**, 2104 (1967); J. O. Jarvie and A. Rauk, *Can. J. Chem.* **52**, 2785 (1974).

(20) H. Basch, *J. Chem. Phys.* **55**, 1700 (1971).

(21) R. Hoffmann, R. Gleiter, and F. B. Mallory, *J. Am. Chem. Soc.* **92**, 1460 (1970).

(22) R. W. Diesen, *J. Chem. Phys.* **39**, 2121 (1963); J. D. Saltzman and E. J. Bair, *J. Chem. Phys.* **41**, 3654 (1964); S. Gordon, W. Mulac, and P. Nangia, *J. Phys. Chem.* **75**, 2087 (1971).

(23) E. M. Evleth, *Chem. Phys. Lett.* **38**, 516 (1976).

(24) R. J. Buenker, S. D. Peyerimhoff, L. C. Allen, and J. L. Whitten, *J. Chem. Phys.* **45**, 2835 (1966).

(25) W. J. Lafferty, A. G. Maki, and T. D. Coyle, *J. Mol. Spectrosc.* **33**, 345 (1970).

(26) W. J. Lafferty and E. K. Plyler, *J. Chem. Phys.* **37**, 2688 (1962); B. P. Stoicheff, *Can. J. Phys.* **40**, 358 (1962); D. W. Lepard, D. M. C. Sweeney, and H. L. Welsh, *Can. J. Phys.* **40**, 1567 (1962).

(27) P. H. Kasai and D. McLeod, Jr., *J. Chem. Phys.* **51**, 1250 (1969).

(28) T. A. Claxton, R. E. Overill, and M. C. R. Symons, *Mol. Phys.* **27**, 701 (1974).

(29) J. W. Rabalais and A. Katrib, *Mol. Phys.* **27**, 923 (1974).

(30) W. A. Lathan, L. A. Curtiss, and J. A. Pople, *Mol. Phys.* **22**, 1081 (1971).

(31) G. W. Mappes, S. A. Fridman, and T. P. Fehlner, *J. Phys. Chem.* **74**, 3307 (1970).

(32) E. V. Waage and B. S. Rabinovitch, *Int. J. Chem. Kinet.* **3**, 105 (1971).

(33) T. P. Fehlner and G. W. Mappes, *J. Phys. Chem.* **73**, 873 (1969).

(34) P. S. Ganguli and H. A. McGee, Jr., *J. Chem. Phys.* **50**, 4658 (1969).

(35) C. Edmiston and P. Lindner, *Int. J. Quantum Chem.* **7**, 309 (1973).

(36) R. Ahlrichs, *Theoret. Chim. Acta* **35**, 59 (1974).

(37) D. S. Marynick, J. H. Hall, Jr., and W. N. Lipscomb, *J. Chem. Phys.* **61**, 5460 (1974).

(38) D. A. Dixon, I. M. Pepperberg, and W. N. Lipscomb, *J. Am. Chem. Soc.* **96**, 1325 (1974).

(39) L. H. Long, *Progr. Inorg. Chem.* **15**, 1 (1972).

(40) J. P. Lowe, *J. Am. Chem. Soc.* **92**, 3799 (1970); **96**, 3759 (1974); J. P. Lowe, *Science* **179**, 527 (1973).

(41) S. Weiss and G. Leroi, *J. Chem. Phys.* **48**, 962 (1968).

(42) D. R. Lide, Jr., *J. Chem. Phys.* **27**, 343 (1957).

(43) E. V. Ivash and D. M. Dennison, *J. Chem. Phys.* **21**, 1804 (1953).

(44) R. Hoffmann and R. A. Olofson, *J. Am. Chem. Soc.* **88**, 943 (1966).

(45) W. J. Hehre and L. Salem, *Chem. Commun.* 754 (1973); W. J. Hehre and J. A. Pople, *J. Am. Chem. Soc.* **97**, 6941 (1975); W. J. Hehre, J. A. Pople, and A. J. P. Devaquet, *J. Am. Chem. Soc.* **98**, 664 (1976).

(46) R. G. Pearson, "Symmetry Rules for Chemical Reactions," p. 255. Wiley, New York, 1976.

7 The full AO basis sets for AB_2 and AB_3

Chapter 3 mentioned a simplified model of the electronic structures of AB_2 and AB_3 halides. In this chapter we use the full, four AOs per atom basis sets to describe in detail the properties of the complete AB_2 and AB_3 classes. These are perhaps the richest classes of small polyatomic molecules. Each class contains a large number of isoelectronic series composed of a wide range of the representative elements. A wealth of detailed experimental information is available for both ground and excited states. Although the AB_2 molecules can only be either linear or bent, several shapes are known to occur within the AB_3 class. Examples of the AB_2 and AB_3 groups have been studied with the most accurate and reliable theoretical methods.

AB_2

SHAPES AND STABILITIES

Table I lists some known AB_2 compounds (*1–82*). The homonuclear species B_3 have been included in the Table but the heteronuclear examples, ABC, with A, B, and C all different have been excluded in an attempt to hold that compilation to a manageable length. However, examples such as BCN, OCS, ClCF, and $BrICl^-$ follow the same rules as their more symmetric relatives. Also omitted are compounds containing atoms of the transition elements and the highly ionic alkaline earth halides. Even with these limitations, Table I contains well over 80 compounds of the better known representative elements.

TABLE I

AB_2 molecules and ions

Valence electrons	AB_2	Shape
10	B_2C	
11	BC_2, BSi_2	
12	C_3, SiC_2, Si_2C	
13	NC_2	
14	CN_2, C_2O	linear
15	BO_2, BS_2, CO_2^+, CS_2^+, N_3, N_2O	
16	CO_2, CS_2, CSe_2, BF_2^+, BO_2^-, CN_2^{2-}, N_3^-, NO_2^+, N_2O, N_2F^+	
17	BF_2, CO_2^-, NO_2, N_3^{2-}, SO_2^+	
18	CF_2, CBr_2, NO_2^-, O_3, SO_2, S_2O, S_3, SeO_2, SiF_2, GeF_2, ClO_2^+	
19	NF_2, NCl_2, PF_2, PCl_2, O_3^-, S_3^-, Se_3^-, SO_2^-, SeO_2^-, NO_2^{2-}, PO_2^{2-}, PS_2^{2-}, AsS_2^{2-}, FO_2, ClO_2	bent
20	OF_2, OCl_2, OBr_2, SF_2, SCl_2, Cl_3^+, Br_3^+, I_3^+, ClF_2^+, $ClBr_2^+$, BrF_2^+, Cl_2F^+, $BrCl_2^+$, ICl_2^+, IBr_2^+, ClO_2^-, BrO_2^-, S_3^{2-}	
21	ClF_2,[a] Cl_3, Br_3	
22	F_3^-, Cl_3^-, Br_3^-, I_3^-, ClF_2^-, $ClBr_2^-$, BrF_2^-, $BrCl_2^-$, BrI_2^-, ICl_2^-, IBr_2^-, KrF_2, XeF_2, $XeCl_2$	linear

[a] See text for mention of this exceptional case.

The AB_2 ground-state shapes follow these simple rules: Molecules with 10 through 16 valence electrons are linear, those with 17 through 20 electrons are bent, and those with 21 or 22 electrons are linear. There seems to be one exception to the linear rule for 21 electron molecules. The infrared spectrum of ClF_2 indicates that this radical is bent with a rather wide valence angle, perhaps 135 to 160° (66). Otherwise, the rules hold remarkably well. Usually, the component atoms of AB_2 molecules are connected such that the central position in both linear and bent shapes is occupied by an atom of the less electronegative element. For example, N_2O is NNO, not NON, and CO_2 is OCO, not COO. NO_2 has a bent symmetric structure ONO. Both ClO_2 (19 electrons) and OCl_2 (20 electrons) are known, each having a bent symmetric structure, as does OBr_2. Apparently, OI_2 has never been prepared. Cl_2Br^- (22 electrons) is linear and symmetrical $ClBrCl^-$, but Br_2Cl^- has the linear and unsymmetrical structure $BrBrCl^-$. Photolysis of ClOCl produces the bent unsymmetrical isomer ClClO, which has been trapped in solid rare gas matrices. Other examples in which a more electronegative atom occupies the central position occur for linear species with fewer than 14 electrons. BC_2 (11 electrons) has been produced in the unsymmetrical form BCC. The isomers CCN and CNC (13 electrons) have been observed as have CNN and NCN (14 electrons). On the basis of ab initio SCF MO calculations (*83*), it is predicted that BCC should be thermodynamically more stable than CBC (11 electrons), CNC more stable than CCN (13 electrons), and NCN more stable than CNN (14 electrons). Thus, the rule that the central position should be occupied by the less electronegative atom has exceptions for bent molecules with 19 or 20 electrons and for linear molecules with fewer than 14 electrons.

MOLECULAR ORBITALS

The AO composition and energy ordering of AB_2 MOs is easier to rationalize in linear geometry. The AO basis set consists of one s and three p AOs on each of three atoms for a total of 12 AOs from which 12 MOs can be formed. The six p AOs that are perpendicular to the molecular axis make up six pi MOs that are grouped in three doubly degenerate sets: $1\pi_u$, $1\pi_g$, and $2\pi_u$. The remaining three p AOs that lie along the molecular axis and the three s AOs combine to form six σ MOs, three of which are σ_g and three are σ_u. The linear AB_2 MOs are shown in Fig. 1. The energies of the orbitals in each group, sigma or pi, increase with the number of nodes that cut across

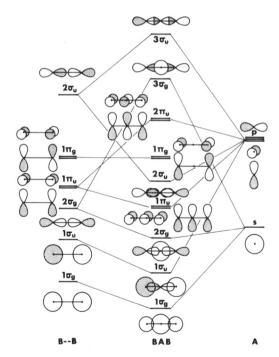

Fig. 1   Formation of MOs for linear BAB from AOs of A and the MOs of a hypothetical B_2 molecule that has the B-B distance of BAB.

the molecular axis. For example, among the sigma orbitals, $1\sigma_g$ (no nodes) has the lowest energy, $1\sigma_u$ (one node) is next, and so on up to $3\sigma_u$ (five nodes). Among the pi MOs the energy increases from $1\pi_u$ (no perpendicular nodes) to $1\pi_g$ (one node) to $2\pi_u$ (two nodes). How the pi energy levels fit in among the sigma levels is harder to rationalize. Figure 1 shows the MOs of linear AB$_2$ as derived by combining the s and p AOs of an isolated A atom with the orbitals of B---B, two B atoms separated by their distance in B-A-B. In constructing Fig. 1, the central atom A was assumed to be somewhat less electronegative than the B atoms. Hence, the s orbital of A is above the energies of the B---B orbitals formed from s AOs, and the p orbitals of A are higher than the center of gravity of the B---B orbitals made from ps. The energy of $1\sigma_g$ (AB$_2$) should be slightly below that of $1\sigma_g$ (B---B) because of the added in-phase overlaps provided through the central A atom s AO. Similarly, the in-phase interaction of s (A) and $1\sigma_u$ (B---B) produces $1\sigma_u$ (AB$_2$) that is lower than its independent component orbitals. There are two pictures one could draw for the double noded MO $2\sigma_g$ (AB$_2$). One possibility is to mix s (A) out-of-phase with $1\sigma_g$ (B---B) to produce an A-B

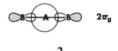

1

antibonding orbital (**1**). The diagram in Fig. 1 emphasizes the bonding interactions that come from mixing s (A) in-phase with $2\sigma_g$ (B---B) (**2**).

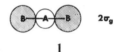

2

A mixture of these two pictures would give a more realistic representation of $2\sigma_g$ (**3**) but the diagram for $2\sigma_g$ (AB$_2$) shown in Fig. 1 is adequate for the

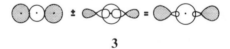

3

purposes here. In-phase and out-of-phase combinations of $1\pi_u$ (B---B) and p (A) make the A-B bonding $1\pi_u$ (AB$_2$) and the A-B antibonding $2\pi_u$ (AB$_2$) MOs that must be, respectively, lower and higher in energy than either $1\pi_u$ (B---B) or p (A) (Fig. 2). The nonbonding $1\pi_g$ (AB$_2$) orbitals receive no AO contributions from the central atom A and, therefore, these MOs have

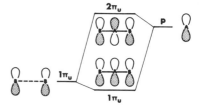

FIG. 2 Detail of the formation $1\pi_u$ and $2\pi_u$ of linear AB_2.

exactly the same energy as $1\pi_g$ (B---B). In-phase and out-of-phase combinations of $2\sigma_u$ (B---B) and the axial p (A) produce, respectively, the bonding $2\sigma_u$ and antibonding $3\sigma_u$ orbitals of AB_2. Surely, the energy of $3\sigma_g$ (AB_2) lies above that of $1\pi_u$ (AB_2) because out-of-phase interactions between the collinear or axial p AOs in the sigma MOs are greater than those among the parallel p AOs in the pi MOs. Similar arguments might place $2\sigma_u$ (AB_2) below $1\pi_u$ (AB_2) instead of above it, as shown in Fig. 1. Justification for the order shown is that there is a larger energy gap between $1\pi_u$ (B---B) and $2\sigma_u$ (B---B) than there is between $1\pi_u$ (B---B) and $2\sigma_g$ (B---B). Thus, the relative meshing of σ and π energy levels for AB_2 as represented in Fig. 1 seems plausible. The order shown is the same as that obtained from extended Hückel calculations for CO_2 and by ab initio SCF MO calculations for NO_2^+ (84). Neither semiempirical nor ab initio MO calculations give this order for all AB_2 systems, however.

The main difference in the energy order of the linear MOs of various AB_2 species is the relative order of $2\sigma_u$ and $1\pi_u$ (AB_2). Several factors influence this order but only one of them will be discussed here with the object being to give an idea of some of the difficulties that arise in trying to develop a qualitative model capable of covering a wide range of molecules. The B---B distance determines the size of the energy gap between $2\sigma_u$ and $1\pi_u$ (B---B). For infinite separation, the $2\sigma_g$, $1\pi_u$, $1\pi_g$, and $2\sigma_u$ (B---B) levels collapse to the 6-fold degenerate p orbitals of two isolated B atoms. If the energy gap between $2\sigma_u$ and $1\pi_u$ (B---B) is small (large separation) and if the energy of p (A) is rather low (small electronegativity difference between A and B), then the stronger interactions among collinear p AOs might push $2\sigma_u$ (AB_2) below $1\pi_u$ (AB_2). What determines the B---B separation? One would expect atoms to be smaller towards the right, through a row of the periodic table as increasing nuclear charge shrinks the radius of the valence orbitals, which would tend to decrease B---B distances. But there is also a swelling effect related to the number of valence electrons and this also increases with nuclear charge. The distance between terminal B atoms will increase as A-B antibonding MOs are filled. The A-B bond distances in Table II show both the shrinkage and the expansion. Fortunately, the energy order chosen in Fig. 1 is adequate for the rationalization of shapes and relative stabilities of the AB_2 triatomics.

TABLE II

Bond distances and bond angles for AB_2 molecules and ions

Valence electrons	AB_2	R(A-B) (Å)	BAB(°)							
16	BeF_2	1.50	180	CO_2	1.162	180	BO_2^-	1.25	180
17	BF_2	1.49	116.5	NO_2	1.193	134.1	CO_2^-	1.24	135.3
18	CF_2	1.30	104.9	O_3	1.27	117.8	NO_2^-	1.236	115.4
19	NF_2	1.353	103.2						
20	OF_2	1.412	103.2						

When the linear $D_{\infty h}$ structure of AB_2 is bent to C_{2v} symmetry, the MO energy changes generally follow the overlap rule (Fig. 3). Consider the $2\pi_u$ orbital, for example. Bending removes the degeneracy of the A-B anti-bonding $2\pi_u$ MOs. The in-phase pi-type B---B overlap in $2b_1$ (C_{2v}) tends to lower slightly the energy of this orbital relative to $2\pi_u$ ($D_{\infty h}$). A much faster energy decrease occurs for $4a_1$ (C_{2v}) as adjacent lobes on parallel p AOs that are A-B antibonding in $2\pi_u$ ($D_{\infty h}$) become less so or perhaps even A-B bonding in bent geometry. The $1\pi_u$ and $2\pi_u$ MOs of linear AB_2 are composed

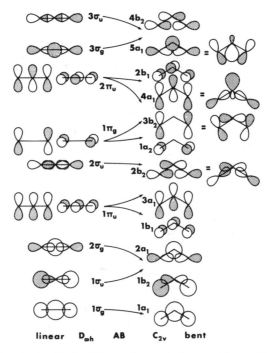

FIG. 3   Qualitative MO correlations for linear and bent AB_2.

solely of *p* AOs parallel to each other and perpendicular to the molecular
axis. Suppose the central and terminal positions are occupied by the same
kind of atom or by atoms of comparable electronegativity. Imagine linear
O_3. The central atom coefficient in $1\pi_u$ would be larger than those for the
terminals, just as the lowest energy wavefunction for the particle in the one-
dimensional box is higher in the middle of the box than at the ends (**4**).

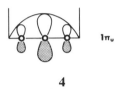

4

MO orthogonality requires that the relative contributions of central atom
and terminal atom *p* AOs in $2\pi_u$ be the opposite of those in $1\pi_u$ (**5**). Now

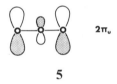

5

suppose the central atom is much less electronegative than the two terminals,
as in BF_2. Perturbation arguments would say that the contributions of the
terminal *p* AOs in $1\pi_u$ would be larger in this example because the fluorine
p AOs have low energy. The coefficient of the high-energy *p* AO from the
central boron would be small in $1\pi_u$ because its energy is high (**6**). The

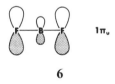

6

relative contributions of AOs in $2\pi_u$ would be just the reverse. This higher
energy MO would be mainly the higher energy *p* AO from the central boron
atom with smaller coefficients for the *p*s on the terminal fluorines (**7**). When

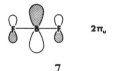

7

the molecule is bent to normal valence angles, the MOs $3a_1$ and $4a_1$ (C_{2v}) retain the balance of p AOs of the related $1\pi_u$ and $2\pi_u$ ($D_{\infty h}$) orbitals. Since the consequences of this composition are important for $4a_1$, diagram **8**

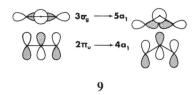

AX large   **AX small**

8

summarizes the results. Here ΔX is the difference in electronegativity between central and terminal atoms, assuming greater electronegativity for the terminals.

In linear geometry $2\pi_u$ and $3\sigma_g$ are rather close in energy. On bending, a component of $2\pi_u$ becomes $4a_1$ (C_{2v}) while $3\sigma_g$ becomes $5a_1$ (C_{2v}) (**9**). Since

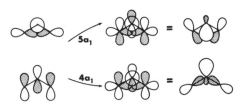

9

$4a_1$ and $5a_1$ are close in energy and are the highest energy MOs of a_1 symmetry, they should mix, as shown in Fig. 4. The extent of this mixing will be greater the smaller the energy gap between $2\pi_u$ and $3\sigma_g$ ($D_{\infty h}$), the unmixable MOs of linear shape. The size of the energy gap is also governed by the electronegativity difference ΔX between A and B. Refer to Fig. 1 to see how $2\pi_u$ and $3\sigma_g$ (AB_2) were formed by out-of-phase overlaps of AOs on A and the $1\pi_u$ and $2\sigma_g$ orbitals of B---B. If ΔX is small or zero as in O_3, these interactions will be large and the stronger sigma-antibonding interactions

FIG. 4   Mixing of the highest energy pair of a_1 orbitals of bent AB_2 produces a hybridlike lobe on the central atom A in $4a_1$.

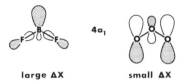

FIG. 5 AO composition extremes in the $4a_1$ MO of AB₂. For BF₂ the difference in electronegativity ΔX between central atom and substituents is large. For O₃, ΔX is zero.

large ΔX   **small ΔX**

in $3\sigma_g$ push this MO above $2\pi_u$ (AB₂), which contains the weaker pi-anti-bonding interactions. If ΔX is large, then the energies of A atom s and p AOs will be high relative to those of B---B. If we raise the A orbitals still higher in energy the perturbation interactions with the B---B orbitals decrease as does the gap between the resulting $3\sigma_g$ and $2\pi_u$ MOs of AB₂. It is conceivable that in some situations the A orbitals could be so high in energy that the perturbation interactions with the B---B orbitals are too weak to push $3\sigma_g$ above $2\pi_u$. The point is that as ΔX increases, $3\sigma_g$ and $2\pi_u$ get closer in energy and the mixing between $4a_1$ and $5a_1$ in bent geometry is more extensive.

Now combine the two electronegativity effects in the same $4a_1$ (C_{2v}) picture. Figure 5 exaggerates the diagrams of $4a_1$ for BF₂ and O₃ in order to emphasize the extremes of AO composition differences due to electronegativity differences. In BF₂ the central atom p AO coefficient of $4a_1$ is large and the mixing with $5a_1$ is extensive yielding a large sp-type hybrid orbital on boron pointing away from the vertex of the valence angle. The small p AOs on the fluorines point toward the boron. Orbital mixing provides extra energy stabilization for $4a_1$ (C_{2v}) and explains the energy increase of $5a_1$ (C_{2v}) from $3\sigma_g$ ($D_{\infty h}$). Overlap considerations alone would have predicted $5a_1$ to fall below $3\sigma_g$. The rise of $5a_1$ has important consequences for the shapes of 21- and 22-electron AB₂ molecules.

Another case of orbital mixing occurs with $3b_2$ and $2b_2$ (C_{2v}) that stem from neighboring $1\pi_g$ and $2\sigma_u$ ($D_{\infty h}$) MOs, respectively (Fig. 6). The $1\pi_g$ and $2\sigma_g$ orbitals are very close together and in some cases their order is reversed. The effect of mixing is to raise the energy of $3b_2$ above that of $1a_2$, which otherwise would appear to remain degenerate by considering the AO compositions of the unmixed MOs. Instead of rising sharply in energy, $2b_2$ is relatively constant for normal valence angles.

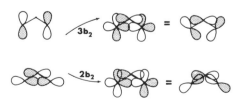

FIG. 6 Mixing of the close lying $2b_2$ and $3b_2$ MOs of bent AB₂.

We have already noted that generally the less electronegative atom occupies the central position in AB_2 molecules. This has been explained as a result of the MO nodal surfaces that pass through or near the center of the molecule, pushing electron density away from the center and towards the terminals. The central atom must be able to give up electron density and the terminals must be able to accept it. Therefore, more electronegative atoms should occupy terminal positions with a less electronegative atom at the center. But the rule seems to be reversed for molecules with 13 electrons or less. These exceptions turn out to be consequences of another aspect of the same nodal surface argument. Among the six MOs of lowest energy, only $1\sigma_u$ and $2\sigma_u$ have nodes that pass through the central atom. Even those two orbitals have central atom axial p AOs that contribute electron density to the central atom. The two $1\pi_u$ MOs and $1\sigma_g$ are the lowest MOs of their respective systems with no nodes cutting perpendicular to the molecular axis. A priori, $1\pi_u$ and $1\sigma_g$ are like the lowest energy wavefunctions for a particle in the one-dimensional box; the wave is higher in the middle than at the ends. Therefore, $1\sigma_g$ and $1\pi_u$ actually enhance electron density at the center. Electronegative atoms prefer locations where the electron density is high (85). Only at the $1\pi_g$ levels, which hold electrons 13–16, do MO nodes completely eliminate AO contributions from the central atom. In molecules with 12 electrons or less the more electronegative element prefers the central position, but the situation reverses as the $1\pi_g$ levels are filled. If the 13-electron radical CNC is more stable than its isomer CCN, it must be because a single electron in $1\pi_g$ cannot overcome the electron density distributions determined by lower occupied orbitals.

In 19- and 20-electron species the preference of electronegative atoms for terminal positions seems to be less strongly held. These are bent molecules. In $3b_2$, bending moves the central atom away from the intersection of nodal surfaces that eliminate the central atom AOs from $1\pi_g$. Mixing of $2b_2$ and $3b_2$ adds a p orbital to the central atom in $3b_2$ and increases the electron density there. The occupied $4a_1$ and $2b_1$ MOs also have central atom AOs that contribute electron density to the central position. If ΔX is not large, the central position might not be too unfavorable as a location for the more electronegative element. The 21- and 22-electron molecules are linear and as $3b_2$ reverts to $1\pi_g$ the electron density at the center decreases and we return to structures in which the central position is occupied by the less electronegative atom.

Now consider the correlation diagram of Fig. 3 in detail. For triatomics with 16 electrons or less, the energies of most of the occupied MOs increase on bending and, therefore, these species are linear. Look at $1b_2$ and $3b_2$ in particular. Only $2a_1$ decreases significantly with bending. Electrons 17 and 18 occupy $2\pi_u$ ($D_{\infty h}$)-$4a_1$ (C_{2v}), which rapidly decreases in energy because of relaxed out-of-phase parallel p overlaps and through MO mixing with $5a_1$. Data in Table II show that experimental bond angles decrease as electrons 17 and 18 are added to $4a_1$. Through the series NO$_2{}^+$ (16 electrons, 180°), NO$_2$ (17 electrons, 134°), NO$_2{}^-$ (18 electrons, 115°), the nuclei are the same and the valence angle decreases as electrons feed into $4a_1$. Electrons 19 and 20 occupy the $2b_1$ (C_{2v}) MO, which also favors bent geometry, if only slightly for normal valence angles. Again, data in Table II show slightly smaller valence angles for 19- and 20-electron examples compared to those for 18-electron species. Electrons 21 and 22 go into $3\sigma_g$ ($D_{\infty h}$)-$5a_1$ (C_{2v}), which favors the linear shape. Except for the 21-electron radical ClF$_2$, the 21- and 22-electron species are linear.

The divergence of $2b_1$ and $4a_1$ (C_{2v}) from $2\pi_u$ ($D_{\infty h}$) calls to mind the similar splitting of π_u into $2a_1$ and $1b_1$ in the AH$_2$ MO system. Both CH$_2$ and NH$_2{}^+$ are bent ground-state triplets with $2a_1$ and $1b_1$ each singly occupied. Both CF$_2$ and NF$_2{}^+$ are ground-state singlets with $4a_1$ doubly occupied and $2b_1$ vacant (86,87).

Details of bond angle and bond distance variations may also be understood from Fig. 2. The $1\pi_g$ MOs are A-B nonbonding. Thus, as electrons 13–16 are added to $1\pi_g$, A-B bond distances should remain relatively constant. Examples are NCN (1.232 Å) versus NCN^{2-} (1.25), CO$_2{}^+$ (1.177) versus CO$_2$ (1.162), CS$_2{}^+$ (1.564) versus CS$_2$ (1.554), and BO$_2$ (1.265) versus BO$_2{}^-$ (1.25). It has been noted (85) that ab initio MO results show that $1\pi_g$ is slightly bonding rather than nonbonding and this accounts for the fact the A-B distances in CS$_2$ and CO$_2$ are slightly shorter (0.01 Å) than those in CS$_2{}^+$ and CO$_2{}^+$. Bond distances in these species are accurately known so the shrinkage is probably real. Other examples that show nearly constant A-B distances are N$_3$ (1.182) and N$_3{}^-$ (1.175). Ab initio SCF MO calculations indicate that N$_3{}^+$ and N$_3$ are linear but unsymmetrical (88a).

Recall that a larger electronegativity difference ΔX between atoms A and B gives added stability to $4a_1$ (C_{2v}), which should favor sharper valence angles. The data in Tables II and III generally support this rule. Notice the decrease in bond angle through the series CF$_2$, SiF$_2$, and GeF$_2$ and through O$_3$, NO$_2{}^-$, and CF$_2$. The SO$_2$ angle, however, is too large to fit the rule for O$_3$, SO$_2$ and SeO$_2$. The A-B antibonding nature of $2b_1$ (C_{2v}) shows up in the lengthening of bonds through the series CF$_2$, NF$_2$, and OF$_2$ as electrons 19 and 20 fill $2b_1$. (See Table II.) Bonds lengthen and valence angles decrease as predicted through the series ClO$_2{}^+$ (18 electrons, 1.31 Å, 122°), ClO$_2$ (19 electrons, 1.475 Å, 117.7°), and ClO$_2{}^-$ (20 electrons, 1.58 Å, 108.4°). Electrons 21 and 22 occupy the $3\sigma_g$ ($D_{\infty h}$)-$5a_1$ (C_{2v}) orbital, which favors linear geometry through mixing with the lower energy $4a_1$. The

TABLE III

Bond distances and bond angles for AB_2 molecules

Valence electrons	AB_2	R(A-B) (Å)	BAB(°)							
18	CF_2	1.30	104.9	SiF_2	1.590	100.8	GeF_2	1.732	97.2
	O_3	1.27	117.8	SO_2	1.432	119	SeO_2	1.609	114
20	OF_3	1.418	103.2	OCl_2	1.70	110.8			
	SF_2	1.589	98.3	SCl_2	2.014	102.8			

larger ΔX, the more the mixing. Therefore, electronegativity considerations would predict ClF_2 as more likely to be linear than the homonuclear triatomics Cl_3 and Br_3. Instead, vibrational spectra indicate that matrix isolated ClF_2 is bent with a rather wide angle while Cl_3 and Br_3 are linear under similar conditions. The $3\sigma_g$-$5a_1$ MO is A-B antibonding and electrons filling this MO should produce further lengthening of A-B bonds. Compare A-B bond distances in bent ICl_2^+ (20 electrons, 2.26 Å) and linear ICl_2^- (22 electrons, 2.55 Å).

EXCITED STATES

Both ground and first excited states of triatomics containing 10 to 15 electrons are linear. Examples of species of known excited state geometry include C_3, CNC, NCN, N_3, BO_2, and CO_2^+. For most 12-, 13-, 14-, and 15-electron species, excitation to the first excited state involves the transfer of an electron from the doubly occupied $2\sigma_u$ MO to a vacancy in $1\pi_g$. For BO_2 and CO_2^+, the excitation has been assigned as being from $1\pi_u$ to $1\pi_g$, which would require a reversal of the order of $1\pi_u$ and $2\sigma_u$ shown in Fig. 1. This could easily be the case since the order of $1\pi_u$ and $2\sigma_u$ is not firmly determined by qualitative considerations.

Although 16-electron molecules and ions are linear in the ground state, their excited states are bent. Excitation is from a filled $1\pi_g$ level to the empty $4a_1$ (C_{2v}) orbital, which favors bent geometry. Bond angles of excited 16-electron species are rather wide, comparable to those for ground-state 17-electron molecules. Examples are CO_2 (122°) and CS_2 (163°). Lowest energy excitation of 17-electron molecules moves an electron from $4a_1$ to $2b_1$. The stabilization of $2b_1$ in bent geometry is small compared to that for $4a_1$. Therefore, NO_2 in the ground-state configuration $(3b_2)^2(4a_1)^1$ is bent at 134°, but in the excited configuration $(3b_2)^2(2b_1)^1$ it is linear. Excitation of an electron from the lower $3b_2$ level to $(3b_2)^1(4a_1)^2$ makes this excited state of NO_2 more strongly bent ($121°$) than the ground state. The electron configuration of the first excited states of 18-electron molecules is $(3b_2)^2(4a_1)^1(2b_1)^1$. Since $2b_1$ is not as effective as $4a_1$ in stabilizing bent shapes, this configuration should give wider valence angles than those in the ground state. Compare ground state SO_2 ($119.5°$) with its first excited state ($126.2°$). Similarly, compare CF_2 in ground ($105°$) and excited ($122.3°$) states. The same trend should be seen in isoelectronic species such as O_3 and SiF_2, but no experimental excited state bond angles are available for these examples. The ground-state configuration of 19-electron molecules is $(3b_2)^2(4a_1)^2(2b_1)^1$. The valence angle of the excited state should be wider whether the excited configuration is $(3b_2)^2(4a_1)^2(5a_1)^1$ or $(3b_2)^2(4a_1)^1(2b_1)^2$. The $(5a_1)^1$ states may be linear. Excited states of 20-electron species may also be linear. No excited state structures of 19- or 20-electron species have been reported. For 21-electron species excited configurations might be $(3b_2)^2(4a_1)^2(2b_1)^1(5a_1)^2$ or $(3b_2)^2(4a_1)^2(2b_1)^2(4b_2)^1$. The $(4b_2)^1$ states should be bent and the $(5a_1)^2$ states should be linear. Again, no experimental excited state structures are known.

HETERONUCLEAR ABC

Essentially the same results and rules that hold for AB_2 also work for the unsymmetrical heteronuclear triatomics ABC. Qualitative MO arguments depend primarily on the positions of MO nodes that are easier to locate if the molecular symmetry is high. But the MOs of unsymmetrical molecules must also have nodes and they are generally not far displaced from positions in the MOs of comparable or related symmetrical examples (85).

CYCLIC STRUCTURES

The relative energy changes of MOs shown in Fig. 3 are for molecules bent at normal valence angles, about $120°$. At smaller angles, some of the energy variations involve dramatic crossings of orbitals of different symmetries. Figure 7 relates MO energy levels of linear ($D_{\infty h}$) and equilateral triangular (D_{3h}) homonuclear B_3. This diagram is based on extended Hückel calculations for O_3. The tie lines between $D_{\infty h}$ and D_{3h} levels are labeled according to the C_{2v} symmetry of the intermediate structure. For the 3-fold symmetric structure it is not convenient to use the same set of p AOs that we used to make MOs for the linear case. Instead, we introduce for each atom the following set: one p orbital perpendicular to the molecular plane, a second lying in the molecular plane and pointing toward the center of the triangle (radial), and the third p in the plane and perpendicular to

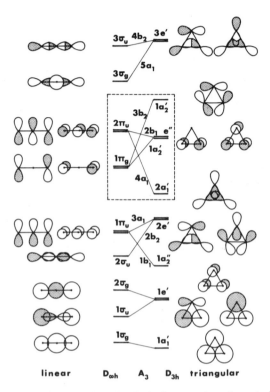

FIG. 7   Qualitative MO correlations for B₃ in linear and equilateral triangular shapes. The energy levels in the dashed rectangle are related to the possible existence of ring-shaped states of B₃ molecules.

the other two (peripheral) (**10**). The order and correlations of the D_{3h} MOs in Fig. 7 can be rationalized. The pi-type orbitals a_2'' and e'' (D_{3h}) correlate in an obvious way with their counterparts in $1\pi_u$, $1\pi_g$, and $2\pi_u$ ($D_{\infty h}$). The all in-phase combination a_2'' must be below $1\pi_u$ because of greater overlap.

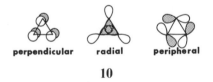

10

Overlap changes also require that the energy of e'' (D_{3h}) be somewhere between the levels of $1\pi_g$ and $2\pi_u$ ($D_{\infty h}$). The head-to-tail out-of-phase arrangement of peripheral p AOs in a_2' (D_{3h}) can be seen to come from $1\pi_g$ ($D_{\infty h}$) through the mixed form of $3b_2$ (C_{2v}) in Fig. 3 (**11**). The energy of

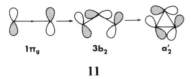

$1\pi_g$ $3b_2$ a_2'

11

a_2' must be relatively high. Comparing the number and strength of the out-of-phase interactions among p AOs should put a_2' (D_{3h}) well above $2\pi_u$ ($D_{\infty h}$) (**12**).The orbital $2a_1'$ (D_{3h}) correlates with $2\pi_u$ ($D_{\infty h}$) through the

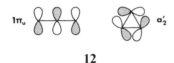

$1\pi_u$ a_2'

12

mixed version of $4a_1$ (C_{2v}) (**13**). Surely, the bonding $2a_1'$ MO must be more

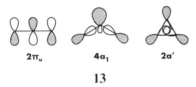

$2\pi_u$ $4a_1$ $2a'$

13

stable than the nonbonding $1\pi_g$ orbitals. Thus, the crossings of several MOs in intermediate C_{2v} geometry are required. Consider carefully the connections between the levels $1\pi_g$, $2\pi_u$ ($D_{\infty h}$) and $2a_1'$, e'', a_2' (D_{3h}) within the rectangle in Fig. 7.

The 14-electron ion N_3^+ has the configuration $(1\pi_g)^2$ in linear symmetric geometry. This configuration is related to $(1a_2)^2$ for the bent C_{2v} shape at normal valence angles. However, as the angle decreases the doubly occupied $1a_2$ orbital rises and crosses the empty $4a_1$ MO that is declining in energy. The crossing of occupied and vacant orbitals of different symmetries signals that the conversion of linear to equilateral triangular shapes is not symmetry-allowed. In configuration interaction terminology, modest C_{2v} deviations from linear geometry are best represented by the configuration $(1a_2)^2$, while small C_{2v} distortions from equilateral triangular geometry are best represented by $(4a_1)^2$. At intermediate geometries neither configuration alone provides an adequate description of the lowest energy electronic state. A better representation of the electronic structure for intermediate geometries would be a linear combination of the two configurations $(1a_2)^2$ and $(4a_1)^2$. Configuration interaction will give a lower energy than will either configuration independently, but at intermediate geometry the energy will probably still be higher than at the extremes of linear and cyclic shape, providing an energy barrier separating minima for linear and cyclic structures. Whether this barrier is large enough to stabilize both shapes is unknown. Although extended Hückel energy levels suggest that the linear and ring structures of N_3^+ have comparable energies, ab initio MO calculations for N_3^+ indicate that the energy of the cyclic structure is 55 to 60 kcal/mole above that of the linear symmetric shape (*88 b, c*). The experimental structure of N_3^+ is linear but the mechanism through which N_3^+ was produced for those studies might be such that formation of the ring state is excluded.

The 16-electron N_3^- has the linear configuration $(1\pi_g)^4$ and is related to $(1a_2)^2(3b_2)^2$ for normal C_{2v} distortions. On bending to the cyclic structure both $1a_2$ and $3b_2$ cross the unoccupied $4a_1$ MO. Once more, the conversion between linear and ring structures is symmetry-forbidden, giving rise to the possibility that both structures might be separated by a stabilizing energy barrier at intermediate geometry. Since the triangular D_{3h} configuration would be the open shell $(2a_1')^2(e'')^2$, it is likely that a C_{2v} form with a valence angle somewhat wider than 60° would be preferred for the N_3^- cyclic state. Ab initio MO calculations indicate that the $(4a_1)^2(1a_2)^2$ configuration of N_3^- has minimum energy with a valence angle near 90° and an energy of about 100 kcal/mole above the linear configuration $(1\pi_g)^4$ (*88d*). The size of the barrier separating the two structures is unknown. Only the linear structure of N_3^- is known from experiment.

The configuration of O_3 (18 electrons) bent at its normal valence angle is $(1a_2)^2(3b_2)^2(4a_1)^2$. At more acute angles the filled $3b_2$ MO crosses the empty $2b_1$, making the $D_{\infty h}$-D_{3h} conversion symmetry-forbidden. Ab initio calculations place the cyclic or ring state of O_3 only 28 to 35 kcal/mole above the normal bent ground state (*89a–e*), only a few kcal/mole above the energy required for dissociation of normal O_3 into $O_2 + O$. Similar calculations for S_3 place the cyclic structure only 5 kcal/mole above the normal bent (117.5°) form (*90*).

Similar orbital crossings among $1a_2$, $3b_2$, $4a_1$, and $2b_1$ occur for the less symmetric AB_2 examples such as NCN (14 electrons), CO_2 and NO_2^+ (16 electrons), and SO_2 and NO_2^- (18 electrons). In these nonhomonuclear cases the degeneracy of $1a_2$ and $2b_1$ as e'' does not occur for cyclic structures. It has been suggested that the high-energy isomer of SO_2, observed in flash photolysis experiments, has an acutely bent or cyclic structure (*89d*). An anomolous form of NO_2^- has been detected in charge transfer and photodetachment experiments. With the thought that this too might have a ring structure, ab initio SCF MO calculations have been done for normal bent, cyclic, as well as bent peroxy NOO^- forms of NO_2^- (*91a*). The results place the cyclic and peroxy forms about 100 and 75 kcal/mole, respectively, above the energy of the normal bent NO_2^- structure. The most likely way by which the peroxy form could relax to the normal structure is through the cyclic structure (**14**). The high energy of the ring form might, therefore,

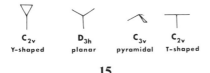

14

serve as a barrier to the rearrangement of peroxy NOO⁻, which could conceivably be formed by a reaction mechanism that leads directly to that metastable state. Similar calculations for the uncharged radical NO_2 suggest that only about 60 kcal/mole separate a peroxy form from the normal bent symmetric structure (*91b*).

Although it is fascinating to speculate about the possible existence of cyclic B_3 and AB_2 structures, there is no firm experimental evidence that any exist and the theoretical work is far from conclusive. However, three small ring compounds of equilateral triangular shape are known. Cyclopropenyl cation (*92a*) $C_3H_3^+$ (14 electrons, isoelectronic with N_3^+) and cyclopropane C_3H_6 and cyclotriazane N_3H_3 (*92b*) (18 electrons, isoelectronic with O_3) have patterns of occupied MOs very similar to that of the cyclic structure in Fig. 7. In these cyclic species, the energy of $2a_1'$ (B_3) would be further stabilized by overlap of AOs of the substituent hydrogens with the radial p AOs of the ring.

AB₃

SHAPES

Table IV lists over 90 compounds of the general formula AB_3 (*93–176*). In the examples discussed here, the central atom A is bound to three ligands or terminals B. In all cases the central atom is less electronegative than the terminals. The AB_3 compounds exhibit four different molecular shapes (**15**).

C_{2v}	D_{3h}	C_{3v}	C_{2v}
Y-shaped	planar	pyramidal	T-shaped

15

Those with 22 valence electrons have planar C_{2v} geometry with one B-A-B angle considerably less than 120° and a weak B-B bond joining the two close ligands to form a three-membered ring. For lack of a better simple descriptive term, we shall call these structures Y shaped. Molecules with 23 or 24

TABLE IV

AB$_3$ molecules and ions

Valence electrons	AB$_3$	Shape
22	CO_3	⎫ Planar,
23	BO_3^{2-}, CO_3^-, NO_3	⎬ Y shaped, C_{2v}
24	BF_3, BCl_3, BBr_3, BI_3, AlF_3, $AlCl_3$, AlI_3, GaF_3, $GaCl_3$, $GaBr_3$, GaI_3, $InCl_3$, $InBr_3$, InI_3, $TlCl_3$, $TlBr_3$, BO_3^{3-}, CO_3^{2-}, CS_3^{2-}, NO_3^-, CF_3^+, CCl_3^+, SO_3, SeO_3	⎫ Planar, ⎬ triangular, ⎭ D_{3h}
25	BF_3^-, CF_3, CCl_3, CBr_3, SiF_3, $SiCl_3$, $GeCl_3$, CO_3^{3-}, NO_3^{2-}, PO_3^{2-}, AsO_3^{2-}, SO_3^-, SeO_3^-, ClO_3, BrO_3	⎫
26	NF_3, NCl_3, PF_3, PCl_3, PBr_3, PI_3, AsF_3, $AsCl_3$, $AsBr_3$, AsI_3, SbF_3, $SbCl_3$, $SbBr_3$, SbI_3, $BiCl_3$, $BiBr_3$, BiI_3, $InCl_3^{2-}$, $InBr_3^{2-}$, InI_3^{2-}, $TlCl_3^{2-}$, $TlBr_3^{2-}$, TlI_3^{2-}, $GeCl_3^-$, $SnCl_3^-$, AsS_3^{3-}, SbS_3^{3-}, SF_3^+, SCl_3^+, SeF_3^+, $SeCl_3^+$, $SeBr_3^+$, $TeCl_3^+$, SO_3^{2-}, SeO_3^{2-}, TeO_3^{2-}, ClO_3^-, BrO_3^-, IO_3^-, XeO_3, TeS_3^{2-}	Nonplanar, pyramidal, C_{3v}
27	SF_3, AsF_3^-, $AsCl_3^-$, BrO_3^{2-}	⎫ distorted T, ⎬ distorted pyramid, C_s
28	ClF_3, BrF_3, $BrCl_3$, $SeCl_3^-$, $SeBr_3^-$, XeF_3^+	⎫ planar, ⎬ T shaped, C_{2v}

valence electrons are planar and equilateral triangular D_{3h}. The 25- and 26-electron examples are nonplanar or pyramidal C_{3v}. Molecules of the 28-electron series are planar with C_{2v} symmetry and have a linear or nearly linear B-A-B backbone that is perpendicular to the third A-B bond. This structure is normally described as T shaped. The 27-electron radicals seem to have structures that are intermediate between pyramidal and T shaped: distorted pyramids or nonplanar distorted Ts.

MOLECULAR ORBITALS

Imagine forming the equilateral triangular D_{3h} structure of AB$_3$ by expanding cyclic B$_3$ enough to accommodate an additional atom A at the center of the triangle and at normal A-B bond distances (16). One can then

16

derive the MOs for AB_3 from those for cyclic B_3 by perturbing them with the new AOs from the central atom A. Since A introduces four new AOs to the basis set, AB_3 has four more MOs than B_3 or a total of 16. The new AOs combine with B_3 MOs as symmetry and energy considerations allow. From the symmetry of the A atom AOs one can deduce that the four new MOs must be a_1' (central atom s), a_2'' (perpendicular p), and e' (in-plane ps).

Figure 8 shows the relationship between the MOs of B_3 and AB_3. Some energy changes are obvious from considerations of overlap changes alone. In the $1a_2'$ orbitals, for example, the head-to-tail out-of-phase overlaps are smaller in AB_3 because the B atoms there are farther apart, giving $1a_2'$ (AB_3) lower energy than $1a_2'$ (B_3). The same explanation holds for the $1e''$ system. The energy gap between $1a_2'$ and $1e''$ must be smaller in AB_3 than in B_3 because the relaxation of the head-to-tail overlaps in $1a_2'$ should be greater than that among the pi-type interactions in $1e''$. Notice that $1a_2'$ is very similar in AO composition to one of the components of $3e'$ (AB_3), the

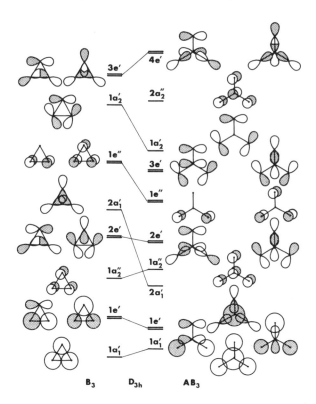

FIG. 8 Formation of MOs for planar triangular AB_3 by expanding those of triangular B_3 and adding the AOs of atom A at the center of the triangle.

main difference being the p AO on A, which can add to $3e'$ but is not allowed in $1a_2'$ by the 3-fold symmetry of that orbital (**17**). Because of the additional

17

A-B overlap in $3e'$, it should have lower energy than $1a_2'$ (AB₃). An important result is that $1e''$, $3e'$, and $1a_2'$ (AB₃) are rather close in energy with $1a_2'$ above the two degenerate pairs.

The energies of $1a_1'$ and $1a_2''$ are slightly higher in AB₃ because B-B in-phase overlaps at bonding distances in B₃ have been replaced by an equal number of A-B overlaps which are not as good. The highest energy AB₃ MO is $3a_1'$ (**18**) (not shown in Fig. 8), the ligand radial p-central s antibonding

18

version of $2a_1'$ (AB₃). Another way to visualize the formation of MOs for AB₃ is to add another B atom to AB₂ bent at 120° (**19**). Such a comparison

19

shows that the $1e''$ (AB₃) orbitals have exactly the same energy as $1a_2$ (AB₂) (**20**). The energy of $3e'$ (AB₃) might be slightly lower than that of $3b_2$ (AB₂)

20

because a ligand p AO adds parallel to and in-phase with one on the central atom (**21**). The new pi-MO $2a_2''$ (AB₃) should be higher than the $2b_1$ (AB₂)

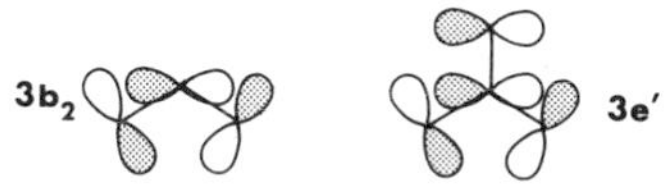

21

orbital from which it can be derived because the new ligand AO must combine out-of-phase with the *p* orbital on the central atom (**22**). In AB_2, $1a_2$

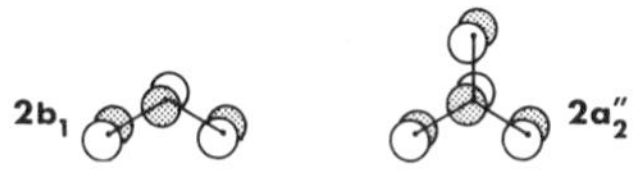

22

is below $3b_2$, with $2b_1$ considerably higher in energy. In AB_3 the related MOs $1e''$ and $3e'$ should be even closer together with $2a_2''$ separated from them by a still larger gap.

The increasing energy order for AB_3, $1e''$, $3e'$, $1a_2'$, $2a_1''$, $4e'$, $3a_1'$, is plausible. The $2a_1''$, $4e'$, and $3a_1'$ MOs are rather close together but separated from the clustered $1e''$, $3e'$, and $1a_2'$ levels below. This ordering of energy levels is supported by extended Hückel calculations for NO_3^- and CO_3^{2-} (*177*) and by CNDO/S calculations for $\dot{N}O_3$ and BF_3 (*178*). However, ab initio SCF MO results for BF_3, CO_3^{2-}, and NO_3^- show $3e'$ below $1e''$ (*179–181*). We will assume the semiempirical order shown in Fig. 8.

Notice that the degenerate $1e''$ pair and $1a_2'$ have nodal surfaces that eliminate any AO contribution from the central atom A. The nodes in these MOs push electron density towards the terminals and ensure that the less electronegative atom occupies the central position.

PLANAR, TRIANGULAR GEOMETRY

The orbitals $3e'$, $1a_2'$, $2a_2''$, and $4e'$ determine the shapes of the AB_3 class. In 24-electron planar, triangular molecules and ions such as BF_3, NO_3^-, CO_3^{2-}, and SO_3, the highest occupied MO is $1a_2'$. This orbital opposes angular changes towards pyramidal (C_{3v}) and T and Y shapes (C_{2v}) because the out-of-phase overlaps among the peripheral *p* AOs on the ligands are at a minimum in planar, equilateral triangular geometry. For example, decreasing one BAB angle towards the Y shape pushes a pair of peripheral *p* AOs together out-of-phase, raising the energy (**23**).

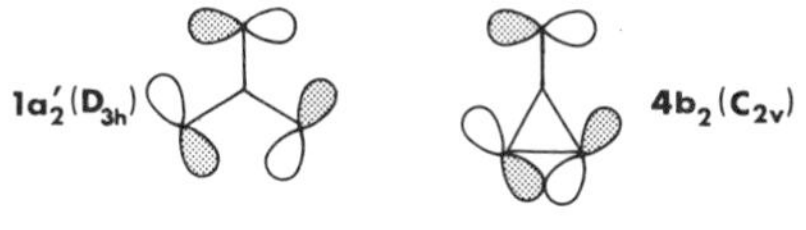

23

THE Y SHAPE

In the 22-electron series, the $1a_2'$ MO is empty and the $3e'$ levels are the highest occupied. If the $3e'$ orbitals are distorted towards the Y shape (C_{2v}), they split into two discrete MOs of a_1 and b_2 symmetry (Fig. 9). From overlap considerations alone, one would expect the a_1 component to stabilize while the b_2 component rises in energy. Thus, the two effects might cancel or, more likely, the out-of-phase character of the b_2 component might dominate to preserve equilateral triangular geometry. But remember that $1a_2'$ lies close above $3e'$ (D_{3h}) and in the Y-shaped structure $1a_2'$ (D_{3h}) becomes $4b_2$ (C_{2v}). Since $3e'$ and $1a_2'$ are close in energy, so are $3b_2$ and $4b_2$ (C_{2v}), which can be expected to mix and diverge, stabilizing $3b_2$. Thus stabilized, $3b_2$ does not block the closing of the unique BAB angle to form the B-B bond of the Y shape as simple overlap considerations might suggest. Instead, the geometry of 22-electron species is controlled by $5a_1$, which closes the three-membered ring of the Y with a B-B bond. The only molecule known to have this structure is CO_3. The analogous CS_3 complex has been proposed as an intermediate in mechanisms of photolytic and shock decompositions of CS_2, reactions very similar to the ones known to produce CO_3 from CO_2.

Can $5a_1$ and $3b_2$ (C_{2v}) produce planar distortions from 3-fold symmetry if $1a_2'$ (D_{3h})-$4b_2$ (C_{2v}) is singly occupied as in the series of 23-electron radicals? Qualitative MO arguments cannot give a firm answer. Semiempirical MO calculations disagree as to whether these distortions lead to energy minima (177,182). The results of ab initio SCF MO calculations for CO_3^- predict planar distortions (183). The structures of 23-electron species are not well determined experimentally, various observations agreeing only that the structures are planar rather than pyramidal. Both CO_3^- and BO_3^{2-} have been observed only in crystal lattices where they are produced by x or γ irradiation. In those environments, electrostatic effects or hydrogen bonding might cause distortions from perfect 3-fold symmetry as often happens to 24-electron ions. NO_3 is known in the gas phase but no definitive experimental structure is available.

How would the qualitative MO explanation of Y-shaped geometry be changed if the order of $1e''$ and $3e'$ in Fig. 8 were reversed as ab initio calcula-

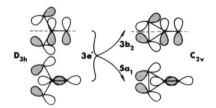

FIG. 9 On distortion to the Y-shaped structure, the $3e'$ orbitals of triangular AB_3 split into $5a_1$ and $3b_2$.

tions indicate? The $1e''$ levels split into a_2 and b_1 orbitals in C_{2v} symmetry. These do not interfere with the a_1 and b_2 MOs from $3e'$ and $1a_2'$. Presumably, the $3e'$ levels would still be close enough to $1a_2'$ for extensive mixing of $3b_2$ and $4b_2$ (C_{2v}) in the Y shape and the explanation remains as before with only the conceptual confusion of intermediate but inert a_2 and b_1 MOs.

PYRAMIDAL SHAPE

In 26-electron molecules the $2a_2''$ (D_{3h}) MO is occupied. Conversion of the planar equilateral triangular D_{3h} structure to the nonplanar or pyramidal C_{3v} shape substantially lowers the energy of $2a_2''$ (D_{3h}) as it is transformed into $4a_1$ (C_{3v}) (**24**). The variation from planar to pyramidal shape moves

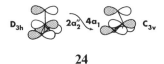

24

parallel p AOs in $2a_2''$ from A-B antibonding positions into an arrangement that is considerably less antibonding or perhaps even A-B bonding. Orbital mixing provides further stabilization of $4a_1$. Above $2a_2''$ (D_{3h}) is $3a_1'$. In pyramidal geometry, $3a_1'$ (D_{3h}) becomes $5a_1$ (C_{3v}). Since $2a_2''$ and $3a_1'$ (D_{3h}) are close together, then $4a_1$ and $5a_1$ (C_{3v}) will also be close. Furthermore, $4a_1$ and $5a_1$ are the highest energy orbitals of a_1 symmetry for pyramidal geometry. Therefore, we can expect extensive $4a_1$, $5a_1$ mixing and considerable lowering of $4a_1$ (C_{3v}), enough so that even a single electron in $4a_1$ produces pyramidal geometry as the structures of the 25-electron series reveal. The 26-electron AB₃ species are even more strongly pyramidal. Notice how the BAB angle closes as electrons 25 and 26 are added to $4a_1$: SeO₃ (24 electrons, 120°), SeO₃⁻ (25 electrons, 112°), SeO₃²⁻ (26 electrons, 100°); GeCl₃ (25 electrons, 111°), GeCl₃⁻ (26 electrons, 96.0°); ClO₃ (25 electrons, 112°), ClO₃⁻ (26 electrons, 106.7°). The angles in the 25-electron radicals are estimates based on electronic structure models and ESR data.

The electronegativity difference ΔX between central atom and substituents affects the valence angle in 25- and 26-electron species in a regular way. First, consider the $1a_2''$ and $2a_2''$ MOs of planar triangular geometry. These are, respectively, the bonding and antibonding pi MOs formed from the parallel p AOs perpendicular to the molecular plane. For small ΔX, the p AO on the central atom makes a larger contribution to $1a_2''$ than do the p AOs from the substituents. In $2a_2''$ the relative contributions are reversed. For large ΔX, the principal components of $1a_2''$ are the ligand or substituent p AOs because these orbitals have low energy. The central atom

p coefficient is relatively smaller. At the $2a_2''$ level the central atom p AO has the larger coefficient because it is the higher energy AO. In pyramidal geometry $2a_1''$ (D_{3h}) becomes $4a_1$ (C_{3v}) and for normal valence angles the relative contributions of central atom and substituent AOs are the same as in $2a_2''$ (D_{3h}) (**25**).

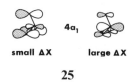

small ΔX   4a₁   large ΔX

25

Above $2a_2''$ (D_{3h}) is the $3a_1'$ (D_{3h}) MO formed by the out-of-phase combination of the central atom s AO with the three radial ps of the substituents. Now the energy gap between $2a_2''$ and $3a_1'$ (D_{3h}) depends on ΔX. Imagine forming planar triangular AB$_3$ by combining the AOs of A with the orbitals of triangular B$_3$ in which the B atoms are separated by their distances in AB$_3$, or in other words, AB$_3$ without the A (Fig. 10). The B$_3$-A combinations are out-of-phase. The sigma-type interactions between a_1'(B$_3$) and s(A) are stronger than the pi-type interactions between a_2''(B$_3$) and p(A), forcing $3a_1'$(AB$_3$) above $2a_2''$(AB$_3$). As the energies of the A atom AOs rise relative to those on B$_3$ (larger ΔX), the size of the perturbation interaction between the A and B$_3$ orbitals is smaller and the energy gap between $2a_2''$ and $3a_1'$(AB$_3$) decreases. For large ΔX the energy difference between $2a_2''$ and $3a_1'$ is small and we can expect greater mixing between $4a_1$ and $5a_1$ (C_{3v}) (Fig. 11). Since this mixing provides additional energy lowering for $4a_1$, larger ΔX makes $4a_1$ more stable and 26-electron molecules more pyramidal.

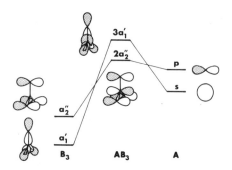

FIG. 10   Formation of antibonding $2a_2''$ and $3a_1'$ MOs of triangular AB$_3$ by combining MOs of B$_3$ with AOs of A. Sigma-type interactions between s(A) and a_1'(B$_3$) place $3a_1'$ above $2a_2''$ which is formed through weaker pi-type interactions of p(A) and a_2''(B$_3$).

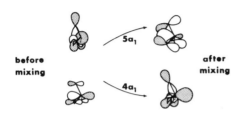

FIG. 11 Mixing of the highest energy pair of a_1 orbitals of pyramidal AB₃ produces a hybridlike lobe on the central atom A in $4a_1$.

There are two electronegativity effects: (1) the relative size of p coefficients for A and B in $4a_1$ and (2) the extent of mixing between $4a_1$ and $5a_1$. Combining these two effects gives the two extreme (large and small ΔX) representations of $4a_1$ shown in Fig. 12.

From the results of ESR experiments on the pyramidal 25-electron radicals, it is possible to estimate the spin densities on the central atom and, further, to divide that spin density between s and p AOs. Those estimates show that central atom spin densities increase as ΔX increases (*127a,130*). The ratios of coefficients p/s decrease as ΔX increases (*127a*) indicating more $4a_1$, $5a_1$ mixing, as summarized in Fig. 12.

Because of the additional stabilization of $4a_1$ (C_{3v}) for larger ΔX, we expect pyramidal molecules to have smaller angles BAB for larger ΔX. There is a wealth of experimental structural data to support this conclusion. Table V contains bond distances and bond angles in the 26-electron pyramidal group V trihalides. This remarkable collection of highly accurate structural data was obtained, mainly by gas phase electron diffraction and microwave spectroscopy, to determine exactly how angles vary with different combinations of central atoms and substituents. (Structures of PI₃ and SbF₃ were obtained by X-ray diffraction studies of crystals.) Two trends in angular variation are apparent: (i) With a particular central atom, the valence angle increases as substituents are replaced by less electronegative ligands (left to right in Table V). (ii) For a given substituent, the valence angle increases as the central atom is replaced by more electronegative atoms (bottom to top). Both trends can be combined into a single rule: As the electronegativity difference ΔX between central atom and ligands increases, the valence angle decreases. The rule has even been incorporated into an empirical formula for calculating valence angles in the group V trihalides as a function of ΔX (*146*).

FIG. 12 AO composition extremes in the $4a_1$ MO of pyramidal AB₃.

  $4a_1$

large ΔX small ΔX

TABLE V

Bond distances and bond angles in group V
trihalides

NF$_3$	NCl$_3$		
1.37 Å	1.76 Å		
102.2°	107.1°		
PF$_3$	PCl$_3$	PBr$_3$	PI$_3$
1.56 Å	2.04 Å	2.22 Å	2.46 Å
96.9°	100.1°	101.0°	102.0°
AsF$_3$	AsCl$_3$	AsBr$_3$	AsI$_3$
1.71 Å	2.16 Å	2.32 Å	2.56 Å
96.2°	98.5°	99.6°	100.2°
SbF$_3$	SbCl$_3$	SbBr$_3$	SbI$_3$
1.92 Å	2.32 Å	2.49 Å	2.72 Å
89°	97.2°	98.2°	99.1°

The electronegativity rule does not apply solely to the group V trihalides. Table VI contains bond distances and bond angles for some trichlorides, including those of Ge, Sn, and Se. Bond angles decrease as ΔX gets larger. The angle in TeCl$_3$$^+$ is unduly small because the Te atom is approximately octahedrally coordinated. The angle shown is for the nonbridging chlorines. Table VII shows that the rule holds for oxygen substituents as well.

T-SHAPED MOLECULES

Electrons 27 and 28 go into the degenerate $4e'$ (D$_{3h}$) orbitals (Fig. 8). Geometrical distortions from 3-fold symmetry can remove the degeneracy of $4e'$ to produce paired electron spins and a closed-shell electronic structure. Under C$_{2v}$ symmetry the e' levels split into a_1 and b_2. By opening one BAB

TABLE VI

26-Electron trichlorides

GeCl$_3$$^-$	AsCl$_3$	SeCl$_3$$^+$
2.27 Å	2.16 Å	2.11 Å
96.0°	98.5°	99.3°
SnCl$_3$$^-$	SbCl$_3$	TeCl$_3$$^+$
2.52 Å	2.32 Å	2.28 Å
89.8°	97.2°	95.0°

TABLE VII

Oxygen substituted pyramidal
structures

SO_3^{2-}	ClO_3^-	
1.52 Å	1.502 Å	
105.7°	106.8°	
SeO_3^{2-}	BrO_3^-	
1.71 Å	1.60 Å	
99.7°	103.0°	
TeO_3^{2-}	IO_3^-	XeO_3
1.88 Å	1.82 Å	1.76 Å
94.6°	99.0°	103.0°

angle to about 180° (T shape), the energy of the a_1 component of $4e'$ (D_{3h}) is lowered because of relaxed out-of-phase overlaps (Fig. 13). Further stabilization due to orbital mixing also occurs. Above $4e'$ (D_{3h}) is $3a_1'$. When $3a_1'$ is opened to the T shape, it has a_1 symmetry and the two highest energy MOs of a_1 symmetry mix. It is the stabilization of the $4e'$ (D_{3h})-$6a_1$(C_{2v}) MO system that makes 28-electron species T shaped. Angular changes to the Y-shaped structure would increase out-of-phase interactions in both a_1 and b_2 (C_{2v}) components of $4e'$ (D_{3h}).

The result of naive mixing of $6a_1$ and $7a_1$ is almost correct. Extended Hückel calculations for ClF_3 show that the $6a_1$ orbital is like a lone pair on the central atom, with a large hydrid-type lobe pointing electron density away from the central atom as shown. But the p AOs at the ends of the T crossbar should point towards the large lone pair lobe (out-of-phase) instead of towards the small nub (in-phase) on the vertical part of the T (**26**). Either picture predicts that the angle between the vertical of the T and

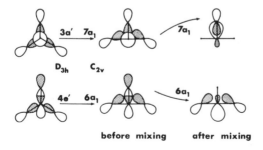

FIG. 13   Mixing of the highest energy pair of a_1 orbitals of T-shaped AB_3.

26

the crossbar might be less than 90°. In the three known experimental structures, these angles vary from 81° for XeF_3^+ to 87.5° for ClF_3. But the extended Hückel result for $6a_1$ gives a clearer picture of why the crossbar A-B bonds are longer (about 0.1 Å) than the vertical A-B bond. The extended Hückel $6a_1$ MO represented above is antibonding with respect to the crossbar substituents while it is nonbonding for the vertical substituent.

REFERENCES

AB₂ Experimental Structural or Synthetic Data

(1) B_2C, BSi_2: G. Verhaegen, F. E. Stafford, and J. Drowart, *J. Chem. Phys.* **40**, 1622 (1964).

(2) BC_2: W. C. Easley and W. Weltner, Jr., *J. Chem. Phys.* **52**, 1489 (1970).

(3) C_3: L. Gausset, G. Herzberg, A. Lagerquist, and B. Rosen, *Astrophys. J.* **142**, 45 (1965).

(4) SiC_2, Si_2C: G. Verhaegen, F. E. Stafford, and J. Drowart, *J. Chem. Phys.* **40**, 1622 (1964).

(5) C_2N: A. J. Merer and D. N. Travis, *Can. J. Phys.* **43**, 1795 (1965); **44**, 353 (1966).

(6)   CNN: D. E. Milligan and M. E. Jacox, *J. Chem. Phys.* **44**, 2850 (1966).

(7)   NCN: G. Herzberg and D. N. Travis, *Can. J. Phys.* **42**, 1658 (1964).

(8) C_2O: M. E. Jacox, D. E. Milligan, N. G. Moll, and W. E. Thompson, *J. Chem. Phys.* **43**, 3734 (1965); C. Devillers and D. A. Ramsay, *Can. J. Chem.* **49**, 2839 (1971).

(9) BO_2: J. W. C. Johns, *Can. J. Phys.* **39**, 1738 (1961).

(10) CO_2^+: J. W. C. Johns, *Can. J. Phys.* **42**, 1004 (1964).

(11) N_3: A. E. Douglas and W. J. Jones, *Can. J. Phys.* **43**, 2216 (1965).

(12) N_2O^+: J. H. Callomon, *Proc. Chem. Soc.* 313 (1959).

(13) BS_2: J. M. Brom, Jr., and W. Weltner, Jr., *J. Mol. Spectrosc.* **45**, 82 (1973).

(14) CS_2^+: W. J. Balfour, *Can. J. Phys.* **54**, 1969 (1976).

(15) CO_2: C.-P. Courtoy, *Can. J. Phys.* **35**, 608 (1957).

(16) BF_2^+: J. C. Kotz, R. J. Vander Zanden, and R. G. Cooks, *Chem. Commun.* 923 (1970).

(17) BO_2^-: C. Calvo and R. Faggiani, *Chem. Commun.* 714 (1974).

(18) NO_2^+: E. Grison, K. Eriks, and J. L. deVries, *Acta Crystallogr.* **3**, 290 (1950).

(19) CN_2^{2-}: Y. Yamamoto, K. Kinoshita, K. Tamaru, and T. Yamanaka, *Bull. Chem. Soc. Jpn.* **31**, 501 (1958).

(20) N_3^-: G. E. Pringle and D. E. Noakes, *Acta Crystallogr.* **B24**, 262 (1968).

(21) N_2O: C. C. Costain, *J. Chem. Phys.* **29**, 864 (1958).

(22) N_2F^+: K. O. Christe, R. D. Wilson, and W. Sawodny, *J. Mol. Struct.* **8**, 245 (1971).

(23) CS_2: C. Jungen, D. N. Malm, and A. J. Merer, *Can. J. Phys.* **51**, 1471 (1973).

(24) CSe_2: T. Wentink, *J. Chem. Phys.* **29**, 188 (1958).

(25) BF_2: W. Nelson and W. Gordy, *J. Chem. Phys.* **51**, 4710 (1969).

(26) CO_2^-: D. W. Ovenall and D. H. Whiffen, *Mol. Phys.* **4**, 135 (1961); S. Schlick, B. L. Silver, and Z. Luz, *J. Chem. Phys.* **54**, 867 (1971); M. E. Jacox and D. E. Milligan, *Chem. Phys. Lett.* **28**, 163 (1974).

(27) NO_2: J. L. Hardwick and J. C. D. Brand, *Chem. Phys. Lett.* **21**, 458 (1973); K.-E. J. Hallin and A. J. Merer, *Can. J. Phys.* **54**, 1157 (1976).

(28) N_3^{2-}: G. W. Neilson and M. C. R. Symons, *J. Chem. Soc. Faraday II* 1772 (1972); T. A. Claxton, R. E. Overill, and M. C. R. Symons, *Mol. Phys.* **26**, 75 (1973).

(29) SO_2^+: G. Schwalbe, S. Schönherr, and W. Windsch, *Chem. Phys. Lett.* **45**, 356 (1977).

(30) CF_2: F. X. Powell and D. R. Lide, Jr., *J. Chem. Phys.* **45**, 1067 (1966); C. W. Mathews, *Can. J. Phys.* **45**, 2355 (1967).

(31) NO_2^-: G. B. Carpenter, *Acta Crystallogr.* **8**, 852 (1955).

(32) O_3: J.-C. Depannemaecker and J. Bellet, *J. Mol. Spectrosc.* **66**, 106 (1977).

(33) SO_2: A. H. Clark and B. Beagley, *Trans. Faraday Soc.* **67**, 2216 (1971); J. D. C. Brand, C. diLauro, and V. T. Jones, *J. Am. Chem. Soc.* **92**, 6095 (1970).

(34) S_2O: E. Tiemann, J. Hoeft, F. J. Lovas, and D. R. Johnson, *J. Chem. Phys.* **60**, 5000 (1974).

(35) ClO_2^+: A. J. Edwards and R. J. C. Sills, *J. Chem. Soc. Dalton* 1726 (1974).

(36) SiF_2: H. Shoji, T. Tanaka, and E. Hirota, *J. Mol. Spectrosc.* **47**, 268 (1973).

(37) S_3: A. G. Hopkins, S.-Y. Tang, and C. W. Brown, *J. Am. Chem. Soc.* **95**, 3486 (1973).

(38) SeO_2: H. Takeo, E. Hirota, and Y. Morino, *J. Mol. Spectrosc.* **34**, 370 (1970); G. W. King, *J. Mol. Spectrosc.* **52**, 154 (1974).

(39) GeF_2: H. Takeo, R. F. Curl, Jr. and P. W. Wilson, *J. Mol. Spectrosc.* **38**, 464 (1971); H. Takeo and R. F. Curl, *J. Mol. Spectrosc.* **43**, 21 (1972).

(40) CBr_2: R. C. Ivey, P. D. Schulze, T. L. Leggett, and D. A. Kohl, *J. Chem. Phys.* **60**, 3174 (1974).

(41) NF_2: R. K. Bohn and S. H. Bauer, *Inorg. Chem.* **6**, 304 (1967); R. D. Brown, F. R. Burden, P. D. Godfrey and I. R. Gillard, *J. Mol. Spectrosc.* **52**, 301 (1974).

(42) NO_2^{2-}: A. Reuveni, P. Poupko, and Z. Luz, *J. Magn. Reson.* **18**, 358 (1975).

(43) FO_2: P. N. Noble and G. C. Pimentel, *J. Chem. Phys.* **44**, 3641 (1966).

(44) O_3^-: M. E. Jacox and D. E. Milligan, *J. Mol. Spectrosc.* **43**, 148 (1972); S. Schlick, *J. Chem. Phys.* **56**, 654 (1972); R. C. Spiker, Jr. and L. Andrews, *J. Chem. Phys.* **59**, 1851 (1973).

(45) PF_2: W. Nelson, G. Jackel, and W. Gordy, *J. Chem. Phys.* **52**, 4572 (1970); M. S. Wei, J. H. Current, and J. Gendell, *J. Chem. Phys.* **52**, 1597 (1970).

(46) ClO_2: A. H. Clark and B. Beagley, *J. Chem. Soc. A* **46** (1970); R. F. Curl *et al.*, *Phys. Rev.* **121**, 1119 (1961).

(47) NCl_2, PCl_2: M. S. Wei, J. H. Current, and J. Gendell, *J. Chem. Phys.* **57**, 2431 (1972).

(48) S_3^-, Se_3^-, SO_2^-, SeO_2^-: J. Schneider, B. Dischler, and A. Räuber, *Phys. Status Solidi* **13**, 141 (1961).

(49) PO_2^{2-}, PS_2^{2-}, AsS_2^{2-}: R. F. Picone, J. B. Raynor, and T. C. Ward, *J. Chem. Soc. Dalton* 392 (1977).

(50) OF_2: L. Pierce, N. DiCianni, and R. H. Jackson, *J. Chem. Phys.* **38**, 730 (1963).

(51) SF_2: D. R. Johnson and F. X. Powell, *Science* **164**, 950 (1969).

(52) OCl_2: J. D. Dunitz and K. Hedberg, *J. Am. Chem. Soc.* **72**, 3108 (1950); R. H. Jackson and D. J. Millen, *Proc. Chem. Soc.* 10 (1959); M. M. Rochkind and G. C. Pimentel, *J. Chem. Phys.* **46**, 4481 (1967).

(53) OBr_2: C. Campbell, J. P. M. Jones, and J. J. Turner, *Chem. Commun.* 888 (1968).

(54) SCl_2: J. T. Murray, W. A. Little, Jr., Q. Williams, and T. L. Weatherly, *J. Chem. Phys.* **65**, 985 (1976); R. W. Davis and M. C. L. Gerry, *J. Mol. Spectrosc.* **65**, 455 (1977).

(55) Cl_3^+: R. J. Gillespie and M. J. Morton, *Inorg. Chem.* **9**, 811 (1970).

(56) Br_3^+: R. J. Gillespie and M. J. Morton, *Inorg. Chem.* **11**, 586 (1972).

(57) I_3^+: D. J. Merryman, P. A. Edwards, J. D. Corbett, and R. E. McCarley, *Chem. Commun.* 779 (1972).

(58) ClF₂⁺: A. J. Edwards and R. J. C. Sills, *J. Chem. Soc. A* 2697 (1970); H. Lynton and
       J. Passmore, *Can. J. Chem.* **49**, 2539 (1971).
(59) BrCl₂⁺, Br₂Cl⁺: W. W. Wilson, B. Landa, and F. Aubke, *Inorg. Nucl. Chem. Lett.* **11**,
       529 (1975).
(60) BrF₂⁺: A. J. Edwards and G. R. Jones, *J. Chem. Soc. A* 1467 (1969).
(61) ICl₂⁺: C. G. Vonk and E. H. Wiebenga, *Acta Crystallogr.* **12**, 859 (1959).
(62) IBr₂⁺: W. W. Wilson and F. Aubke, *Inorg. Chem.* **13**, 326 (1974).
(63) S₃²⁻: S. Yamaoka, J. T. Lemley, J. M. Jenks, and H. Steinfink, *Inorg. Chem.* **14**, 129
       (1975).
(64) ClO₂⁻: V. Tazzoli, V. Riganti, G. Giuseppetti, and A. Coda, *Acta Crystallogr.* **B31**,
       1032 (1975); C. Tarimci, E. Schempp, and S. C. Chang, *Acta Crystallogr.* **B31**, 2146
       (1975); C. Tarimci, R. D. Rosenstein, and E. Schempp, *Acta Crystallogr.* **B32**, 610
       (1976).
(65) BrO₂⁻: M.-T. LeBihan, B. Gurtner, and A. Kalt, *Bull. Soc. Fr. Miner. Crystallogr.*
       98, 223 (1975).
(66) ClF₂: G. Mamantov, D. G. Vickroy, E. J. Vasini, T. Maekawa, and M. C. Moulton,
       Inorg. Nucl. Chem. Lett. **6**, 701 (1970); G. Mamantov, E. J. Vasini, M. C. Moulton,
       D. G. Vickroy, and T. Maekawa, *J. Chem. Phys.* **54**, 3419 (1971); E. S. Prochaska and
       L. Andrews, *Inorg. Chem.* **16**, 339 (1977).
(67) Cl₃: L. Y. Nelson and G. C. Pimentel, *J. Chem. Phys.* **47**, 3671 (1967).
(68) Br₃: D. H. Boal and G. A. Ozin, *J. Chem. Phys.* **55**, 3598 (1971).
(69) F₃⁻: B. S. Ault and L. Andrews, *J. Am. Chem. Soc.* **98**, 1591 (1976).
(70) Cl₃⁻: J. C. Evans and G. Y.-S. Lo, *J. Chem. Phys.* **44**, 3638 (1966).
(71) Br₃⁻: S. L. Lawton, E. R. McAfee, J. E. Benson, and R. A. Jacobson, *Inorg. Chem.*
       12, 2939 (1973).
(72) I₃⁻: J. Runsink, S. Swen-Walstra, and T. Migchelsen, *Acta Crystallogr.* **B28**, 1331
       (1972).
(73) ClF₂⁻: K. O. Christe and J. P. Guertin, *Inorg. Chem.* **4**, 1785 (1965); K. O. Christe,
       W. Sawodny, and J. P. Guertin, *Inorg. Chem.* **6**, 1159 (1967).
(74) ClBr₂⁻: J. C. Evans and G. Y.-S. Lo, *J. Chem. Phys.* **45**, 1069 (1966).
(75) BrF₂⁻: T. urles, L. A. Quarterman, and H. H. Hyman, *J. Inorg. Nucl. Chem.* **35**,
       668 (1973).
(76) BrCl₂⁻: W. Gabes and H. Gerding, *J. Mol. Struct.* **14**, 267 (1972).
(77) BrI₂⁻: A. G. Maki and R. Forneris, *Spectrochim. Acta* **23A**, 867 (1967).
(78) ICl₂⁻: F. vanBolhuis and P. A. Tucker, *Acta Crystallogr.* **B29**, 2613 (1973).
(79) IBr₂⁻: J. E. Davies and E. K. Nunn, *Chem. Commun.* 1374 (1969).
(80) KrF₂: H. H. Claassen, G. L. Goodman, J. G. Malm, and F. Schreiner, *J. Chem. Phys.*
       42, 1229 (1965).
(81) XeF₂: S. Reichman and F. Schreiner, *J. Chem. Phys.* **51**, 2355 (1969).
(82) XeCl₂: W. F. Howard, Jr. and L. Andrews, *J. Am. Chem. Soc.* **96**, 7864 (1974).

AB₂ Calculated Structures

(83) BCC, CBC, CCN, CNC, CNN, NCN: C. Thomson, *J. Chem. Phys.* **58**, 216, 841 (1973).
(84) NO₂⁺, NO₂⁻: G. V. Pfeiffer and L. C. Allen, *J. Chem. Phys.* **51**, 190 (1969).
(85) NNO: S. D. Peyerimhoff and R. J. Buenker, *J. Chem. Phys.* **49**, 2473 (1968).
(86) CF₂: J. F. Harrison, *J. Am. Chem. Soc.* **93**, 4112 (1971).
(87) NF₂⁺: J. F. Harrison and C. W. Eakers, *J. Am. Chem. Soc.* **95**, 3467 (1973).
(88) N₃⁺, N₃, N₃⁻: (a) T. W. Archibald and J. R. Sabin, *J. Chem. Phys.* **55**, 1821 (1971);
       (b) J. Wright, *J. Am. Chem. Soc.* **96**, 4753 (1974); (c) T. J. Venanzi and J. M. Schulman,

Mol. Phys. **30**, 281 (1975); (d) S. D. Peyerimhoff and R. J. Buenker, *J. Chem. Phys.* **47**, 1953 (1967).

(89) O_3: (a) S. D. Peyerimhoff and R. J. Buenker, *J. Chem. Phys.* **47**, 1953 (1967); (b) P. J. Hay, T. H. Dunning, and W. A. Goddard, *Chem. Phys. Lett.* **23**, 457 (1973); (c) S. Shih, R. J. Buenker, and S. D. Peyerimhoff, *Chem. Phys. Lett.* **28**, 463 (1974); (d) E. F. Hayes and G. V. Pfeiffer, *J. Am. Chem. Soc.* **90**, 4773 (1968); (e) L. B. Harding and W. A. Goddard, III, *J. Chem. Phys.* **67**, 2377 (1977).

(90) S_3: N. R. Carlsen and H. F. Schaefer, III, *Chem. Phys. Lett.* **48**, 390 (1977).

(91) NOO, NOO$^-$: (a) P. K. Pearson, H. F. Schaefer, III, J. H. Richardson, L. M. Stephenson, and J. I. Brauman, *J. Am. Chem. Soc.* **96**, 6778 (1974); (b) H. F. Schaefer, III, C. F. Bender, and J. H. Richardson, *J. Phys. Chem.* **80**, 2035 (1976).

(92) (a) R. Breslow, J. T. Groves, and G. Ryan, *J. Am. Chem. Soc.* **89**, 5048 (1967); (b) Y. Kim, J. W. Gilje, and K. Seff, *J. Am. Chem. Soc.* **99**, 7057 (1977).

AB_3 *Experimental Structural or Synthetic Data*

(93) CO_3: N. G. Moll, D. R. Clutter, and W. E. Thompson, *J. Chem. Phys.* **45**, 4469 (1966); M. E. Jacox and D. E. Milligan, *J. Chem. Phys.* **54**, 919 (1971).

(94) BO_2^{2-}: P. C. Taylor, D. L. Griscom, and P. J. Bray, *J. Chem. Phys.* **54**, 748 (1971).

(95) CO_3^-: S. A. Marshall and R. A. Serway, *J. Chem. Phys.* **46**, 1949 (1967).

(96) NO_3: E. Haydon and E. Saito, *J. Chem. Phys.* **43**, 4314 (1965); T. W. Martin, L. L. Swift, and J. H. Venable, Jr., *J. Chem. Phys.* **52**, 2138 (1970); D. E. Wood and G. P. Lozos, *J. Chem. Phys.* **64**, 546 (1976).

(97) BF_3: P. A. Freedman and W. J. Jones, *J. Mol. Spectrosc.* **54**, 182 (1975); K. Kuchitsu and S. Konaka, *J. Chem. Phys.* **45**, 4342 (1966); S. Konaka, Y. Murata, K. Kuchitsu, and Y. Morino, *Bull. Chem. Soc. Jpn.* **39**, 1134 (1966).

(98) BCl_3: S. Konaka, Y. Murata, K. Kuchitsu, and Y. Morino, *Bull. Chem. Soc. Jpn.* **39**, 1134 (1966).

(99) BBr_3: S. Konaka, T. Ito, and Y. Morino, *Bull. Chem. Soc. Jpn.* **39**, 1146 (1966).

(100) BI_3: H. Kakubari, S. Konaka, and M. Kimura, *Bull. Chem. Soc. Jpn.* **47**, 2337 (1974).

(101) AlF_3: P. A. Akishin, N. G. Rambidi, and E. Z. Zasorin, *Kristallografia* **4**, 186 (1959).

(102) $AlCl_3$: P. A. Akishin, N. G. Rambidi, and E. Z. Zasorin, *Kristallografia* **4**, 186 (1959); I. Hargittai and M. Hargittai, *J. Chem. Phys.* **60**, 2563 (1974); I. R. Beattie, H. E. Blayden, and J. S. Ogden, *J. Chem. Phys.* **64**, 909 (1976).

(103) AlI_3: Q. Shen, *Diss. Abstr. Int. B* **34**, 3735 (1974).

(104) GaF_3: F. M. Brewer, G. Garton, and D. M. L. Goodgame, *J. Inorg. Nucl. Chem.* **9**, 56 (1959).

(105) $GaCl_3$: R. G. S. Pong, R. A. Stachnik, A. E. Shirk, and J. S. Shirk, *J. Chem. Phys.* **63**, 1525 (1975); M. C. Drake and G. M. Rosenblatt, *J. Chem. Phys.* **65**, 4067 (1976).

(106) $GaBr_3$: Q. Shen, *Diss. Abstr. Int. B* **34**, 3735 (1974); R. G. S. Pong, R. A. Stachnik, A. E. Shirk, and J. S. Shirk, *J. Chem. Phys.* **63**, 1525 (1975).

(107) GaI_3: Y. Morino, T. Ukaji, and T. Ito, *Bull. Chem. Soc. Jpn.* **39**, 71, 191 (1966); M. C. Drake and G. M. Rosenblatt, *J. Chem. Phys.* **65**, 4067 (1976).

(108) $InCl_3$: I. R. Beattie, H. E. Blayden, S. M. Hall, S. N. Jenny, and J. S. Ogden, *J. Chem. Soc. Dalton*, 666 (1976); R. G. S. Pong, A. E. Shirk, and J. S. Shirk, *J. Mol. Spectrosc.* **66**, 35 (1977).

(109) $InBr_3$, InI_3: J. J. Habeeb and D. G. Tuck, *Chem. Commun.* 808 (1975).

(110) $TlCl_3$, $TlBr_3$: Z. V. Zvonkova, *Zh. Fiz. Khim. SSSR* **30**, 340 (1956).

(111) BO_3^{3-}: A. Vegas, F. H. Cano, and S. Garcia-Blanco, *Acta Crystallogr.* **B31**, 1416 (1975).

(112) CO_3^{2-}: R. L. Sass, R. Vidale, and J. Donohue, *Acta Crystallogr.* **10**, 567 (1957).

(113)　CS$_3^{2-}$: E. Philippot and O. Lindqvist, *Rev. Chem. Miner.* **8**, 491 (1971).

(114)　NO$_3^-$: G. L. Paul and A. W. Pryor, *Acta Crystallogr.* **B27**, 2700 (1972).

(115)　ClF$_3^+$: C. Lifshitz and W. A. Chupka, *J. Chem. Phys.* **47**, 3439 (1967).

(116)　CCl$_3^+$: M. E. Jacox and D. E. Milligan, *J. Chem. Phys.* **54**, 3935 (1971).

(117)　SO$_3$: A. Kaldor and A. G. Maki, *J. Mol. Struct.* **15**, 123 (1973).

(118)　SeO$_3$: F. C. Mijlhoff, *Rec. Trav. Chim.* **84**, 74 (1965).

(119)　BF$_3^-$: R. L. Hudson and F. Williams, *J. Chem. Phys.* **65**, 3381 (1976).

(120)　CF$_3$: R. W. Fessenden and R. H. Schuler, *J. Chem. Phys.* **43**, 2704 (1965).

(121)　CCl$_3$: L. Andrews, *J. Chem. Phys.* **48**, 972 (1968); G. Maass, A. K. Maltser, and J. L. Margrave, *J. Inorg. Nucl. Chem.* **35**, 1945 (1973).

(122)　CBr$_3$: R. C. Ivey, P. D. Schulze, T. L. Leggett, and D. A. Kohl, *J. Chem. Phys.* **60**, 3174 (1974).

(123)　SiF$_3$: D. E. Milligan, M. E. Jacox, and W. A. Guillory, *J. Chem. Phys.* **49**, 5330 (1968); M. V. Merritt and R. W. Fessenden, *J. Chem. Phys.* **56**, 2353 (1972).

(124)　SiCl$_3$: M. E. Jacox and D. E. Milligan, *J. Chem. Phys.* **49**, 3130 (1968); R. V. Lloyd and M. T. Rogers, *J. Am. Chem. Soc.* **95**, 2459 (1973).

(125)　GeCl$_3$: W. A. Guillory and C. E. Smith, *J. Chem. Phys.* **53**, 1661 (1970); R. V. Lloyd and M. T. Rogers, *J. Am. Chem. Soc.* **95**, 2459 (1973).

(126)　CO$_3^{3-}$: S. A. Marshall and R. A. Serway, *J. Chem. Phys.* **46**, 1949 (1967); R. S. Eachus and M. C. R. Symons, *J. Chem. Soc. A* 790 (1968).

(127)　NO$_3^{2-}$: (a) R. S. Eachus and M. C. R. Symons, *J. Chem. Soc. A* 790 (1968); (b) M. C. R. Symons, *J. Chem. Soc. A* 1998 (1970).

(128)　PO$_3^{2-}$: M. C. R. Symons, *J. Chem. Soc. A* 1998 (1970); S. Schlick, B. L. Silver, and Z. Luz, *J. Chem. Phys.* **52**, 1232 (1970).

(129)　SO$_3^-$: G. W. Chantry, A. Horsefield, J. R. Morton, J. R. Rowlands, and D. H. Whiffen, *Mol. Phys.* **5**, 233 (1962).

(130)　SeO$_3^-$: P. W. Atkins, M. C. R. Symons, and H. W. Wardale, *J. Chem. Soc.* 5215 (1964).

(131)　AsO$_3^{2-}$: W. C. Lin and C. A. McDowell, *Mol. Phys.* **7**, 223 (1963/64).

(132)　ClO$_3$: P. W. Atkins, J. A. Brivati, N. Keen, M. C. R. Symons and P. A. Trevalion, *J. Chem. Soc.* 4785 (1962); J. B. Bates and H. D. Stidham, *Chem. Phys. Lett.* **37**, 25 (1976).

(133)　BrO$_3$: J. R. Byberg and B. S. Kirkegaard, *J. Chem. Phys.* **60**, 2594 (1974).

(134)　NF$_3$: M. Otake, C. Matsumura, and Y. Morino, *J. Mol. Spectrosc.* **28**, 316 (1968).

(135)　NCl$_3$: H. B. Burgi, D. Stedman, and L. S. Bartell, *J. Mol. Struct.* **10**, 31 (1971); G. Cazzoli, *J. Mol. Spectrosc.* **53**, 37 (1974).

(136)　PF$_3$: Y. Morino, K. Kuchitsu, and T. Moritani, *Inorg. Chem.* **8**, 867 (1969); E. Hirota and Y. Morino, *J. Mol. Spectrosc.* **33**, 460 (1970); Y. Kawashima and A. P. Cox, *J. Mol. Spectrosc.* **65**, 319 (1977).

(137)　PCl$_3$: P. Kisliuk and C. H. Townes, *J. Chem. Phys.* **18**, 1109 (1950); K. Hedberg and M. Iwasaki, *J. Chem. Phys.* **36**, 589 (1962); G. Cazzoli, *J. Mol. Spectrosc.* **53**, 37 (1974).

(138)　PBr$_3$: K. Kushitsu, T. Shibata, A. Yokozeki, and C. Matsumura, *Inorg. Chem.* **10**, 2584 (1971).

(139)　PI$_3$: E. T. Lance, J. M. Haschke, and D. R. Peacor, *Inorg. Chem.* **15**, 780 (1976).

(140)　AsF$_3$: F. B. Clippard, Jr., and L. S. Bartell, *Inorg. Chem.* **9**, 805 (1970); S. Konaka and M. Kimura, *Bull. Chem. Soc. Jpn.* **43**, 1693 (1970).

(141)　AsCl$_3$: P. Kisliuk and C. H. Townes, *J. Chem. Phys.* **18**, 1109 (1950); S. Konaka and M. Kimura, *Bull. Chem. Soc. Jpn.* **43**, 1693 (1970).

(142)　AsBr$_3$: S. Samdal, D. M. Barnhart, and K. Hedberg, *J. Mol. Struct.* **35**, 67 (1976); A. G. Robiette, *J. Mol. Struct.* **35**, 81 (1976).

(143) AsI$_3$: Y. Morino, T. Ukaji, and T. Ito, *Bull. Chem. Soc. Jpn.* **39**, 71, 191 (1966).

(144) SbF$_3$: A. J. Edwards, *J. Chem. Soc. A* 2751 (1970).

(145) SbCl$_3$: S. Konaka and M. Kimura, *Bull. Chem. Soc. Jpn.* **46**, 404 (1973); G. Cazzoli and W. Caminati, *J. Mol. Spectrosc.* **62**, 1 (1976).

(146) SbBr$_3$: S. Konaka and M. Kimura, *Bull. Chem. Soc. Jpn.* **46**, 413 (1973).

(147) SbI$_3$: A. Almenningen and T. Bjorvatten, *Acta Chem. Scand.* **17**, 2573 (1963).

(148) BiCl$_3$; BiBr$_3$: H. A. Skinner and L. E. Sutton, *Trans. Faraday Soc.* **36**, 681 (1940).

(149) BiI$_3$: T. R. Manley and D. A. Williams, *Spectrochim. Acta* **21**, 1773 (1965).

(150) InCl$_3^{2-}$, InBr$_3^{2-}$, InI$_3^{2-}$: G. Contreras and D. G. Tuck, *Chem. Commun.* 1552 (1971).

(151) TlCl$_3^{2-}$, TlBr$_3^{2-}$, TlI$_3^{2-}$: D. C. Luehrs, *J. Inorg. Nucl. Chem.* **31**, 3517 (1969).

(152) GeCl$_3^-$: S. Fregerslev and S. E. Rasmussen, *Acta Chem. Scand.* **22**, 2541 (1968).

(153) SnCl$_3^-$: F. R. Poulsen and S. E. Rasmussen, *Acta Chem. Scand.* **24**, 150 (1970).

(154) AsS$_3^{3-}$, SbS$_3^{3-}$: D. Harker, *J. Chem. Phys.* **4**, 381 (1936).

(155) SF$_3^+$: D. D. Gibler, C. J. Adams, M. Fischer, A. Zalkin, and N. Bartlett, *Inorg. Chem.* **11**, 2325 (1972).

(156) SCl$_3^+$: H. E. Doorenbos, J. C. Evans, and R. O. Kagel, *J. Phys. Chem.* **74**, 3385 (1970).

(157) SeF$_3^+$: A. J. Edwards and G. R. Jones, *J. Chem. Soc. A* 1491 (1970).

(158) SeCl$_3^+$: H. Gerding and D.-J. Stufkens, *Rev. Chim. Miner.* **6**, 795 (1969); B. A. Stork-Blaisse and C. Romers, *Acta Crystallogr.* **B27**, 386 (1971); A. Bali and K. C. Malhotra,. *J. Inorg. Nucl. Chem.* **39**, 957 (1977).

(159) SeBr$_3^+$: A. Bali and K. C. Malhotra, *J. Inorg. Nucl. Chem.* **39**, 957 (1977).

(160) TeCl$_3^+$: H. Gerding and D.-J. Stufkens, *Rev. Chim. Miner.* **6**, 795 (1969); B. Krebs, B. Buss, and D. Altena, *Z. Anorg. Allg. Chem.* **386**, 257 (1971).

(161) SO$_3^{2-}$: L. Niinistö and L. O. Larsson, *Acta Crystallogr.* **B29**, 623 (1973); I. Hjerten and B. Nyberg, *Acta Chem. Scand.* **27**, 345 (1973); L. O. Larsson and L. Niinistö, *Acta Chem. Scand.* **27**, 859 (1973).

(162) SeO$_3^{2-}$: T. Asai and R. Kiriyama, *Bull. Chem. Soc. Jpn.* **46**, 2395 (1973).

(163) TeO$_3^{2-}$: O. Lindqvist, *Acta Chem. Scand.* **26**, 1423 (1972).

(164) ClO$_3^-$: S. K. Sikka, S. N. Momin, H. Rajagopal, and R. Chidambaran, *J. Chem. Phys.* **48**, 1882 (1968); M. E. Burke-Laing and K. N. Trueblood, *Acta Crystallogr.* **B33**, 2698 (1977).

(165) BrO$_3^-$: G. Gjörnlund, *Acta Chem. Scand.* **25**, 1645 (1971).

(166) IO$_3^-$: N. W. Alcock, *Acta Crystallogr.* **B28**, 2783 (1972); P. D. Cradwick and A. S. deEndredy, *J. Chem. Soc. Dalton* 1926 (1975).

(167) XeO$_3$: D. H. Templeton, A. Zalkin, J. D. Forrester, and S. M. Williamson, *J. Am. Chem. Soc.* **85**, 817 (1963).

(168) TeS$_3^{2-}$: J.-C. Jumas, M. Ribes, M. Maurin, and E. Philippot, *Acta Crystallogr.* **B32**, 444 (1976).

(169) SF$_3$: A. J. Colussi, J. R. Morton and K. F. Preston, *J. Chem. Phys.* **61**, 1247 (1974).

(170) AsF$_3^-$, AsCl$_3^-$: S. Subramanian and M. T. Rogers, *J. Chem. Phys.* **57**, 4582 (1972).

(171) BrO$_3^{2-}$: J. R. Byberg and B. S. Kirkegaard, *J. Chem. Phys.* **60**, 2594 (1974).

(172) ClF$_3$: R. D. Burbank and F. N. Bensey, *J. Chem. Phys.* **21**, 602 (1953); D. F. Smith, *J. Chem. Phys.* **21**, 609 (1953).

(173) BrF$_3$: R. D. Burbank and F. N. Bensey, *J. Chem. Phys.* **27**, 982 (1957).

(174) BrCl$_3$: L. Y. Nelson and G. C. Pimentel, *Inorg. Chem.* **7**, 1695 (1968).

(175) SeCl$_3^-$, SeBr$_3^-$: K. J. Wynne and J. Golen, *Inorg. Chem.* **13**, 185 (1974).

(176) XeF$_3^+$: D. E. McKee, C. J. Adams, A. Zalkin, and N. Bartlett, *Chem. Commun.* 26 (1973); P. Boldrini, R. J. Gillespie, P. R. Ireland, and G. J. Schrobilgen, *Inorg. Chem.* **13**, 1690 (1974).

General References

(177)   B. M. Gimarc and T. S. Chou, *J. Chem. Phys.* **49**, 4043 (1968).

(178)   H. J. Maria, J. R. McDonald, and S. P. McGlynn, *J. Am. Chem. Soc.* **95**, 1050 (1973).

(179)   T. E. H. Walker and J. A. Horsley, *Mol. Phys.* **21**, 939 (1971).

(180)   J. F. Wyatt, I. H. Hillier, V. R. Saunders, J. A. Connor, and M. Barber, *J. Chem. Phys.* **54**, 5311 (1971).

(181)   J. A. Connor, I. H. Hillier, V. R. Saunders, and M. Barber, *Mol. Phys.* **23**, 81 (1972).

(182)   J. F. Olsen and L. Burnelle, *J. Am. Chem. Soc.* **92**, 3659 (1970).

(183)   S. P. So, *J. Chem. Soc. Faraday II*, **72**, 646 (1976).

8 Strengths of single bonds in symmetric dimers

The A_2B_2 molecules and ions with 18 and 26 valence electrons, the A_2B_4 species with 34 and 38 electrons, and the 50-electron A_2B_6 series all have structures in which single bonds link AB_n monomers in symmetric dimers $B_nA\text{-}AB_n$. Trends in the A-A bond strength can be understood within the MO framework as a function of the electronegativity difference ΔX between central atoms A and substituent or terminal atoms B. Electronegativity differences can also account for the nonexistence of some A_2B_{2n} compounds or rationalize their preference for less symmetric structures.

A_2B_4

Table I contains the A-A bond distances in some A_2B_4 molecules and ions with 34 valence electrons. Two trends in these data are worthy of note. First, the A-A bond distance is longer for N_2O_4 than for $C_2O_4^{2-}$. One might expect the N-N distance to be shorter as a larger nuclear charge contracts the $2s$ and $2p$ valence AOs of the central atoms. The value $2r_A$, twice the covalent single bond radius of A, is smaller for N_2O_4 than for $C_2O_4^{2-}$. The quantity $\Delta r = R(A\text{-}A) - 2r_A$ measures the deviation from the atomic radius additivity rule. One could hope that the radius additivity concept might apply through the series B_2F_4, $C_2O_4^{2-}$, and N_2O_4 since these species have the same set of MOs occupied by the same number of valence electrons. Compare the N-N single bond in N_2O_4 (1.78 Å) [1,2] with N-N single bonds

TABLE I

A_2B_4. 34 Valence electron series

	B_2F_4	$C_2O_4^{2-}$	N_2O_4	B_2Cl_4
$R(A\text{-}A)$ (Å)	1.72	1.57	1.78	1.75
$2r_A$ (Å)	1.60	1.54	1.48	1.60
Δr (Å)	0.12	0.03	0.30	0.15
$D(B_2A\text{-}AB_2)$ (kcal/mole)	88 ± 15		13	83 ± 4
ΔX	2.0	0.9	0.4	1.1

in N_2F_4 (1.492 Å) (3,4) and N_2H_4 (1.449 Å) (5). There may be some uncertainty about how long a normal B-B bond should be, but clearly the C-C
bond in $C_2O_4^{2-}$ (1.57 Å) (6) is rather long compared with single bonds in
ethane (1.536 Å) and diamond (1.545 Å) (7). In summary, the central bond
in $C_2O_4^{2-}$ is a little longer than expected and that in N_2O_4 is much longer.
A second point of interest is the large difference in A-A bond strength
between B_2F_4 and N_2O_4. The B-B bond in B_2F_4 is much stronger than the
N-N bond in N_2O_4.

There is a third important anomaly in the A_2B_4 series. N_2O_4 is planar
D_{2h} with a barrier to rotation about the N-N bond that is large (2 to 3
kcal/mole) (8,9) considering the long and weak (13 kcal/mole) (2,10) N-N
bond. The Raman spectra of gaseous and solid B_2F_4 are consistent with a
planar molecular structure in these states (11). Recent electron diffraction
studies also indicate that gaseous B_2F_4 has a planar D_{2h} structure (12).
Ab initio SCF MO calculations favor staggered geometry for B_2F_4 (13–15).
All studies agree that the barrier to internal rotation about the B-B bond
is small (0.5 to 1.8 kcal/mole). X-ray diffraction experiments show that B_2F_4
is planar in the solid (16). B_2Cl_4 (17–22) and B_2Br_4 (23) are both staggered
in the gas phase. Crystalline $C_2O_4^{2-}$ is usually planar, but in some cases
the two CO_2 groups may be twisted out of coplanarity by as much as 26°
(24). Vibrational spectra suggest that $C_2O_4^{2-}$ may be staggered in aqueous
solutions (25,26). Thus, molecules and ions of the 34-electron A_2B_4 series
may violate a Walsh-type rule that isoelectronic species should have the
same molecular conformation (27). The conformational preferences are also
functions of the electronegativity difference between central atoms and
substituents.

The 34 valence electrons of the A_2B_4 series occupy 17 molecular orbitals.
The highest occupied MO is the sigma-bonding MO $4a_g$. Among the 16
MOs of lower energy there are equal members of A-A bonding and A-A
antibonding MOs. Thus, there is no net A-A bonding for A_2B_4 molecules
with 32 valence electrons and, indeed, none are known with a direct A-A

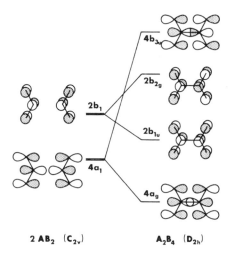

2 AB₂ (C₂ᵥ)   **A₂B₄ (D₂ₕ)**

FIG. 1   MOs of two AB_2 monomers combine to form bonding and antibonding MOs of the A_2B_4 dimer. [Reprinted with permission from Gimarc, *J. Am. Chem. Soc.* **100**, 1996 (1978). Copyright by the American Chemical Society.]

bond. Instead, the 32-electron molecules Be_2F_4 (*28*) and Be_2Cl_4 (*29*) apparently have planar D_{2h} structures that involve bridging atoms. If two 16-electron AB_2 monomers were brought together in coplanar fashion in an attempt to form an A-A link, not only would the A-A interactions be repulsive but also the net out-of-phase interactions between B substituents on opposed monomers would be minimized if the monomers were rotated to the staggered D_{2d} conformation.

In the 34-electron planar D_{2h} system, the highest occupied MO, $4a_g$, is responsible for the symmetric A-A bonding of two AB_2 monomers. The $4a_g$ MO can be formed by the in-phase combination of two $4a_1$ MOs from a pair of bent (C_{2v}) AB_2 monomers, as shown in Fig. 1. The dependence of the AO composition of $4a_1$ (AB_2) on the electronegativity difference ΔX between substituents and central atoms was described in detail in Chapter 7, using orbital mixing and perturbation arguments. For BF_2 (large ΔX), $4a_1$ has a large hybrid orbital on boron pointing away from the vertex of the FBF valence angle (**1**). The small p AOs on the fluorines point towards

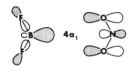

1

the rear nub of the boron hybrid. In NO_2 (small ΔX), the $4a_1$ orbital has a small p AO on the central N and large parallel p AOs on the terminal Os. The $4a_1$ MOs of two AB_2 monomers combine in-phase to form the bonding $4a_g$ orbital of A_2B_4 (2). The large hybrid orbitals pointing towards each

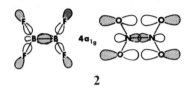

2

other from the two borons form a strong B-B bond in B_2F_4. The small p AOs on the fluorines cannot interact strongly with those on the other end of the B-B bond. Net end-end antibonding interactions from the lower occupied MOs may dominate to give B_2F_4 the staggered conformation or at least act to lower the rotational barrier of a planar conformation. In N_2O_4 the small nitrogen p AOs on the N-N axis overlap to form a weak dimer. On the other hand, large p AOs on the oxygens of opposite monomers overlap in-phase to overcome the antibonding interactions of the lower orbitals and to stabilize a planar conformation for N_2O_4.

The rule of weaker central A-A bonds with decreasing electronegativity difference between A and B predicts that the B-B bond in B_2Cl_4 should be longer and weaker than that in B_2F_4. The experimental bond distances (18–21) and dissociation energies suggest that this might be true, although the large uncertainties in those values make the comparison questionable. However, the larger size of the terminal atom AOs could also lengthen and weaken the B-B bond in B_2Cl_4 by increasing the out-of-phase interactions in the lower occupied valence MOs. In the series B_2F_4, $C_2O_4^{2-}$, and N_2O_4, all central atoms are from the same row of the periodic table and all terminals are also from the same row. For the comparison of B_2F_4 and B_2Cl_4 the terminals are from different rows. The higher principal quantum number of the valence AOs on chlorine gives those orbitals a larger effective radius and, therefore, the net antibonding or repulsive interactions from the 16 lower occupied MOs would be larger, lengthening and weakening the B-B bond of B_2Cl_4 compared to that of B_2F_4. These larger antibonding interactions are apparently responsible for the increase in the barrier to internal rotation from B_2F_4 (0.5 to 1.8 kcal/mole) (12,14) to B_2Cl_4 (1.5 to 2.5 kcal/mole) (14,19–21) to B_2Br_4 (2.11 to 3.75 kcal/mole) (23).

The electronegativity rule can also be used to rationalize the non-existence of the thiooxalate anion $C_2S_4^{2-}$. For this ion ΔX is smaller than that for N_2O_4 and one would predict an even weaker dimer bond. The larger sulfur atoms could also weaken the bond.

The orbital $4a_g$ is formed by the in-phase combination of two $4a_1$ MOs from the AB_2 monomers. The corresponding out-of-phase combination $4b_{3u}$ is much higher in energy. Between $4a_g$ and $4b_{3u}$ lie $2b_{1u}$ and $2b_{2g}$, which are, respectively, the bonding and antibonding combinations of the $2b_1$ MOs of the AB_2 monomers. Refer once more to Fig. 1. The $2b_{1u}$ MO makes the pi-bond in 36-electron molecules such as C_2F_4. In 38-electron molecules the antibonding $2b_{2g}$ MO is occupied, canceling the pi-bond and yielding another A-A single bonded series that includes N_2F_4 and $S_2O_4^{2-}$. If 40 electrons were to occupy this MO system, the antibonding $4b_{3u}$ orbital would be filled, canceling the sigma A-A link of $4a_g$. The 40-electron molecule S_2F_4 is known (30), but it has the unsymmetrical structure F_3S-SF rather than the symmetrical B_2AAB_2 pattern that is ruled out by MO considerations.

The A_2B_4 class contains two series of singly bonded dimers. Molecules in the 38-electrons series have nonplanar trans C_{2h} conformations with possible rotational isomers of gauche C_2 symmetry. An explanation of the rotational conformations is the subject of another story but it is easy to see why the B_2A-AB_2 system prefers to be nonplanar with 38 electrons. Out-of-plane folding changes the $2b_{1u}$ and $2b_{2g}$ MOs from A-B antibonding orbitals of planar geometry to A-B bonding orbitals for pyramidal geometry about A. Diagram **3** shows this transformation for $2b_{2g}$. Folding the A_2B_4

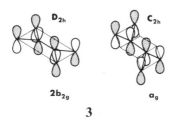

3

structure to trans shape changes $4a_g$ (D_{2h}) into $4a_g$ (C_{2h}) (**4**). Thus, the $4a_g$ MOs that are responsible for A-A bonding in both the 34 (D_{2h})- and 38 (C_{2h})-electron series are directly related through AO composition, and the same electronegativity rule applies.

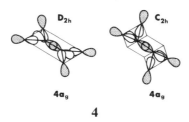

4

TABLE II

A_2B_4. 38 Valence electron series

	N_2F_4	P_2F_4	$S_2O_4^{2-}$
$R(A-A)$ (Å)	1.49	2.281	2.389
$2r_A$ (Å)	1.48	2.20	2.08
Δr (Å)	0.01	0.08	0.31
$D(B_2A-AB_2)$ (kcal/mole)	20.4	57 ± 10	21
ΔX	0.9	1.8	0.8

Table II contains data for some 38-electron A_2B_4 molecules. As expected, the A-A bond in $S_2O_4^{2-}$ (31,32) is longer and weaker than that in P_2F_4 (33,34). There is some question about the electronegativity rule in the diphosphines. Estimates of P-P bond energies seem to increase through the series P_2F_4 (57 ± 10 kcal/mole), P_2Cl_4 (62 kcal/mole), P_2I_4 (71 to 80 kcal/mole) (34), but the uncertainties in these estimates are quite large. The P-P bond distance in P_2F_4 (2.28 Å) (34) is longer than that in P_2I_4 (2.21 Å) (35). On the other hand, the calculated P-P stretching force constant for P_2Cl_4 is greater than that for P_2I_4 (36). The compounds X_2O_4 (X = halogen) are known. Because fluorine is the most electronegative element, one would not expect it to occupy the central position where MO nodes accumulate to force electron density away from the center and toward less electronegative terminals. Instead, F_2O_4 has the peroxidetype structure FOOOOF (37). The electronegativities of Cl and O are close ($\Delta X = 0.3$) and one would not expect a strong Cl-Cl bond in Cl_2O_4. Instead, this compound has the unsymmetrical structure O_3Cl-O-Cl (38). Symmetrical Br_2O_4 is more likely since ΔX is larger (0.5) and, indeed, vibrational spectra favor the structure O_2Br-BrO_2 (39). The Mössbauer spectrum of I_2O_4 in frozen H_2SO_4 solution suggests $O_2IO^- \cdots IO^+$ (40), which is related to the structure of the polymeric solid (41). Our electronegativity rule predicts a symmetrical structure because ΔX is larger (0.8) than for Br_2O_4. Finally, N_2O_4 and N_2F_4 offer one comparison between the 34- and the 38-electron series. The larger electronegativity difference in N_2F_4 makes its N-N bond shorter and stronger (42) than that in N_2O_4.

A_2B_6

Table III lists some known ethanelike A_2B_6 molecules and ions containing 50 valence electrons. These electrons occupy 25 MOs, the highest of which is the A-A sigma bonding $4a_{1g}$ MO. Below $4a_{1g}$ lie 24 MOs among which are equal numbers of A-A bonding and A-A antibonding orbitals. The net effect of these 24 underlying MOs is therefore end-end repulsive,

TABLE III

A_2B_6. 50 Valence electron series

	Si_2F_6	$P_2O_6^{4-}$	$S_2O_6^{2-}$	Si_2Cl_6	$P_2S_6^{4-}$	$P_2Se_6^{4-}$
$R(A\text{-}A)$ (Å)	2.32	2.17	2.15	2.32	2.20	2.24
$2r_A$ (Å)	2.34	2.20	2.08	2.34	2.20	2.20
Δr (Å)	-0.02	-0.03	$+0.07$	-0.02	0	$+0.04$
ΔX	2.1	1.2	0.8	1.2	0.4	0.4

giving 50-electron molecules their staggered D_{3d} conformation. Although there are many 48-electron A_2B_6 molecules, none has a direct A-A bond between the two AB_3 monomers. Instead, the 48-electron molecules have diboranelike structures with a pair of bridging atoms. Examples are Al_2I_6 (43), Ga_2Cl_6 (44), and $Ge_2S_6^{4-}$ (45).

The $4a_{1g}$ MO of A_2B_6 forms the bond between the AB_3 monomers. This MO is the in-phase combination of singly occupied $4a_1$ MOs of the two AB_3 monomers (5). The $4a_1$ MO gives 25- and 26-electron AB_3 molecules

5

their pyramidal (C_{3v}) shape. The change from planar to pyramidal shape moves parallel p AOs from A-B antibonding positions into A-B bonding positions, just like that for $4a_1$ (AB_2, C_{2v}). The electronegativity difference ΔX between substituents and central atoms determines the AO composition of $4a_1$ (AB_3) as described in Chapter 7. The AO compositions of $4a_1$ (AB_2) and $4a_1$ (AB_3) are related. Therefore, there is a direct relationship between $4a_{1g}$ (A_2B_6) and $4a_g$ (A_2B_4) and the electronegativity rule we found for 34- and 38-electron A_2B_4 molecules holds for the 50-electron A_2B_6 series as well.

The only A-A bond energy comparison for the 50-electron A_2B_6 series is between C_2F_6 (96.5 kcal/mole) (46) and C_2Cl_6 (72.4 kcal/mole) (47). However, the substituents have valence AOs of different principal quantum number and the larger substituent AOs could also act to weaken the C-C bond. Larger substituent orbitals are presumably responsible for the increasing barrier to internal rotation from C_2F_6 (3.91 kcal/mole) (48) to C_2Cl_6 (10.8 to 17.5 kcal/mole) (49) to C_2Br_6 (13 to 50 kcal/mole?) (50). Most comparisons of A-A bond distances are for systems in which variations might also be due to both the electronegativity rule and substituent orbital size. The two series $Ga_2Cl_6^{2-}$ (2.390 Å) (51), $Ga_2Br_6^{2-}$ (2.419) (52) and

$P_2O_6^{4-}$ (2.17) (53), $P_2S_6^{2-}$ (2.20), $P_2Se_6^{4-}$ (2.24) (54) show clear lengthening of the central A-A bond with increasing substituent orbital size and decreasing ΔX. The difference of 0.02 Å between observed C-C bond distances in C_2F_6 (1.545 Å) (55) and C_2Cl_6 (1.564) (56) is surprisingly small considering the 24 kcal/mole difference in dissociation energies D (X_3C-CX_3). The reported C-C bond distance in C_2Br_6 (1.526 Å) (57) is shorter than that in either C_2F_6 or C_2Cl_6. However, these central bond distances are difficult to measure accurately by X-ray or electron diffraction techniques because the two carbons are virtually surrounded and, therefore, obscured by six strongly scattering halogen atoms.

Another measure for the comparison of A-A bond strengths are calculated A-A stretching force constants. A pattern of decreasing Si-Si stretching force constants has been observed in the Si_2X_6 series: Si_2F_6 (2.4 mdyn/Å), Si_2Cl_6 (2.4 mdyn/Å), Si_2Br_6 (2.1 mdyn/Å), Si_2I_6 (1.9 mdyn/Å) (58). The In-In stretching constants in $In_2X_6^{2-}$ seem to follow a similar trend: $In_2Cl_6^{2-}$ (0.63 ± 0.08 mdyn/Å), $In_2Br_6^{2-}$ (0.69 ± 0.06 mdyn/Å), $In_2I_6^{2-}$ (0.24 ± 0.01 mdyn/Å) (59). One might hope to find the same pattern for the C-C stretches in the perhaloethanes C_2X_6. The vibrational spectral data are available, but apparently force constants for this series have never been calculated without the assumption of equal C-C stretching force constants throughout (60).

In the series Si_2F_6 (61), $P_2O_6^{4-}$, $S_2O_6^{2-}$ (62), and Cl_2O_6, all central atoms come from the same row of the periodic table and all substituents B come from the same row. Therefore, it should be possible to isolate the effect of electronegativity differences on A-A bond strengths and distances. Table III compares the observed bond distances R(A-A) with twice the covalent radii $2r_A$. The difference $\Delta r = R(A-A) - 2r_A$ is a measure of the deviation of R(A-A) from the radius additivity rule. For Si_2F_6 and $P_2O_6^{4-}$, Δr is negligible, the negative value perhaps indicating that the A-A distance is even a little shorter than predicted by the radius additivity rule. The small value and negative sign of Δr for $P_2O_6^{4-}$ also eases concern that the large charge on this ion might produce an expansion of the ion and a lengthening of the P-P bond. The value of $\Delta r = +0.07$ Å for $S_2O_6^{2-}$ signals a lengthening of the S-S bond. The compound Cl_2O_6 has long been known (63). ΔX is very small. Recent spectroscopic measurements of the liquid indicate that the ClO_3 dimer is not symmetric or ethanelike, but rather it has the unsymmetrical structure O_3Cl-O-ClO_2 (64).

A_2B_2

Both the 18- and 26-electron series of the BAAB class have A-A single bonds. Members of the 18-electron series are C_2N_2, C_2P_2, B_2O_2, and B_2S_2. C_2N_2 has a linear symmetric structure N≡C—C≡N and bond distances

TABLE IV

S_2X_2. 26 Valence electron series

	S_2F_2	S_2Cl_2	S_2Br_2	S_4^{2-}
$R(S-S)$ (Å)	1.89	1.93	1.98	2.07
Δr (Å) $(2r_S = 2.08$ Å)	−0.19	−0.15	−0.10	−0.01
ΔX	1.4	0.6	0.4	0

are known accurately from X-ray diffraction measurements (65). The infrared spectrum of B_2O_2 is consistent with a linear symmetric molecule OBBO (66) and this is the structure assumed here although thermodynamic data have recently been used to argue in favor of a bent unsymmetrical molecule (67). The 26-electron series includes O_2F_2 (68), S_2F_2 (69), S_2Cl_2 (70), S_2Br_2 (71), Se_2Cl_2 (72), and Se_2Br_2 (72). These molecules are known to have chainlike, nonplanar, gauche C_2 shapes like hydrogen peroxide. The molecule S_2I_2 is stable only at low temperatures. Its structure is unknown (73).

The members of the 26-electron series seem to follow the electronegativity rule for A-A bond strengths developed for single bonded A_2B_4 and A_2B_6 systems. The data in Table IV show that the S-S bond grows longer through the series S_2F_2, S_2Cl_2, S_2Br_2 as ΔX decreases. For O_2F_2 the dissociation energy D (FO-OF) is estimated to be 62.1 kcal/mole (74). The Cl_2O_2 dimer, however, is very weakly bound and it possibly has the unsymmetrical structure ClO-ClO (75). The 26-electron homonuclear ions S_4^{2-} and I_4^{2+} are weakly bound as the electronegativity rule would predict. The dissociation energy of I_4^{2+} into $2I_2^+$ is estimated to be around 10 kcal/mole (76). In liquid solutions S_4^{2-} is in equilibrium with the radical ions $2S_2^-$ (77). The X-ray crystal structure of BaS_4 has been determined (78) and the central S-S bond in the S_4^{2-} ion is long (2.07 Å) compared to that in S_2Cl_2 (1.93 Å), a molecule containing terminal atoms of greater electronegativity but with valence AOs of the same principal quantum number as the sulfur terminals in the homonuclear ion S_4^{2-}. In contrast, the 18-electron series does not appear to follow the electronegativity rule. It turns out that exactly the same principles apply, but the result is a reversal of the effect for the linear 18-electron series. The electronegativities of C and N are fairly close yet the central bond in $N\equiv C-C\equiv N$ has a dissociation energy of 128 kcal/mole (79), very strong and short (1.38 Å) compared to the C-C bond in C_2F_6 (96.5 kcal/mole, 1.545 Å). Apparently the central bond in the OB-BO structure is weaker than that in NC-CN despite the larger electronegativity difference ΔX between B and O than between C and N. Estimates of D(OB-BO) range from 100 to 120 kcal/mole (66,80). The B-B distance in B_2O_2 is unknown. From the heats of atomization of CP and C_2P_2 (81) one

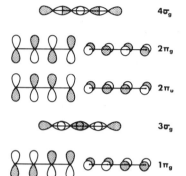

$4\sigma_g$

$2\pi_g$

$2\pi_u$

$3\sigma_g$

$1\pi_g$

FIG. 2   Some of the valence MOs of linear A_2B_2. The $3\sigma_u$ MO that lies between $2\pi_g$ and $4\sigma_g$ has been omitted here because it plays no part in our discussion. [Reprinted with permission from Gimarc, *J. Am. Chem. Soc.* **100**, 1996 (1978). Copyright by the American Chemical Society.]

can calculate $D(\text{PC-CP}) = 149 \pm 8$ kcal/mole. In this case, ΔX is smaller than for C_2N_2 and the carbon-carbon bond is stronger.

Figure 2 shows some of the higher energy valence MOs for linear BAAB. In the 18-electron BAAB series the highest occupied MO is the $3\sigma_g$ orbital formed by the in-phase overlap of the colinear p AOs on the four atoms BAAB. Below $3\sigma_g$, 16 electrons occupy eight MOs of which half are A-A bonding and half are A-A antibonding. Thus, it is $3\sigma_g$ that provides the A-A single bond which links the two AB monomers in the linear A_2B_2 dimer. As just stated, our model would not account for 16-electron A_2B_2 species with a direct A-A bond. In fact, C_4 (16 electrons) has been detected in the vapor over graphite at high temperatures. The abundance of C_4 is less than that of either C_3 or C_5 (*82*). Now $3\sigma_g$ and $1\pi_g$ are close in energy and their order is poorly determined by qualitative considerations. They might be reversed in the case of C_4. If $3\sigma_g$ were below $1\pi_g$, then two C_2 monomers would be joined by the bonding $3\sigma_g$ MO and C_4 would be an open shell or triplet state because $1\pi_g$ would be only half filled, a configuration supported by MO calculations (*83*).

For large ΔX the p AOs from the terminal B atoms would make the larger contribution to $3\sigma_g$ because the B atom ps have lower energy than those from the central atoms A. For small ΔX the central A-A p contributions are large, again by analogy with the lowest energy wave function for the particle in the one-dimensional box. The two extreme cases are illustrated in **6**. The

$3\sigma_g$

large ΔX small ΔX

6

larger the p coefficients on A, the stronger the A-A bond. Thus, for larger ΔX A-A bonds should become weaker for the linear 18-electron series.

Moving from the 18-electron A_2B_2 series to the 26-electron series, eight electrons are added to two sets of doubly degenerate MOs, $2\pi_u$ and $2\pi_g$. The A-A bonding character of the $2\pi_u$ orbitals is canceled by the antibonding nature of $2\pi_g$ and, therefore, the 26-electron series should have A-A single bonds, which they do. If that single bond were due to the same $3\sigma_g$ MO responsible for bonding in the 18-electron series, then one would expect weaker A-A bonds for larger ΔX, a trend opposite to that observed for the 26-electron series. The resolution of this dilemma is based on the fact that the 26-electron molecules are not linear. Although the 26-electron species are actually nonplanar or gauche C_2, we assume planar, trans C_{2h} geometry for representational convenience. If $3\sigma_g$ were bent trans, one would expect its energy to rise because colinear p AO overlaps between atoms A and B would be reduced. Above $2\pi_g$ is $4\sigma_g$, the A-B antibonding combination of colinear p AOs related to $3\sigma_g$. On bending, the energy of $4\sigma_g$ ought to decrease because A-B out-of-phase overlaps are relaxed. Both $3\sigma_g$ and $4\sigma_g$ produce MOs of a_g symmetry in trans geometry. Bending produces a reversal of order of the two a_g MOs derived from $3\sigma_g$ and $4\sigma_g$. That this must be so can be seen from Fig. 3 in which these same MOs are formed through in-phase combinations of the appropriate MOs of two AB monomers to form directly the trans-bent dimer. Figure 3 shows that by the in-phase combination of two A-B bonding 1π MOs of trans associated AB monomers one can form an a_g dimer orbital that has the same phase relationships among component p AOs as those in $4\sigma_g$ of the linear structure. Similarly, the in-phase combination of two antibonding 2π MOs of AB monomers leads to the $3\sigma_g$ MO of the linear dimer. The dashed lines connecting 1π (AB) and $4\sigma_g$ (A₂B₂) and between 2π (AB) and $3\sigma_g$ (A₂B₂) represent intended correlations that are not allowed by the noncrossing rule. The actual picture is even more complicated because of the mixing of other MOs of a_g symmetry since there are six a_g MOs that can be made from our AO basis set. The a_g MO that maintains net A-A bonding in the nonlinear 26-electron series is

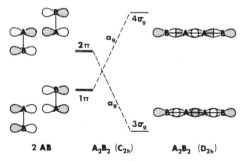

2 AB   **A₂B₂ (C₂ₕ)**   **A₂B₂ (D₂ₕ)**

FIG. 3   The intended correlations of pi MOs of trans oriented AB monomers with the sigma MOs of the linear dimer A₂B₂. [Reprinted with permission from Gimarc, *J. Am. Chem. Soc.* **100**, 1996 (1978). Copyright by the American Chemical Society.]

that made by combining the two 2π (AB) MOs. It is, after all, related to $3\sigma_g$ (A_2B_2) by AO phases. But Fig. 3 shows how the a_g MO related to $4\sigma_g$ (A_2B_2) slips underneath in nonlinear geometry.

Now consider how the 1π and 2π MOs of the AB monomer are influenced by ΔX. As for AB_2 and AB_3, we are concerned with a pair of monomer MOs formed from parallel p AOs. If ΔX is zero (homonuclear case), the two p AO coefficients must be equal in both 1π and in the higher energy 2π. If ΔX is large, then the p AO contribution of B in 1π will be larger than that of A because the AO energies of B are lower. Conversely, the p AO contribution from A will be larger in the higher energy combination 2π. Diagram 7 portrays the extreme cases. Since the size of the A atom p con-

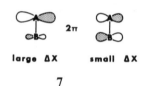

large ΔX small ΔX

7

tribution determines the strength of the A-A bond, then large ΔX systems will form strong A-A bonds, just as they did in the A_2B_4 and A_2B_6 classes. There is a difference, however. For $\Delta X = 0$, the A and B p AO coefficients in 2π must be equal; small ΔX does not diminish the A atom p contribution below that of the B atom, a consequence that can occur in the comparable MOs of AB_2 and AB_3. The requirement that the p contribution from A to 2π cannot be smaller than that from B, therefore, limits the weakening suffered by the BA-AB bond for small ΔX cases and accounts for the existence of homonuclear ions S_4^{2-} and I_4^{2+}, which have no counterparts in the A_2B_4 and A_2B_6 classes.

REFERENCES

(1) B. W. McClelland, G. Gundersen, and K. Hedberg, *J. Chem. Phys.* **56**, 4541 (1972).
(2) Q. Shen, *Diss. Abstr. Int. B* **34**, 3735 (1974).
(3) M. J. Cardillo and S. H. Bauer, *Inorg. Chem.* **8**, 2086 (1969).
(4) M. M. Gilbert, G. Gundersen, and K. Hedberg, *J. Chem. Phys.* **56**, 1691 (1972).
(5) Y. Morino, T. Iijima, and Y. Murata, *Bull. Chem. Soc. Jpn.* **33**, 46 (1960).
(6) B. F. Pederson, *Acta Chem. Scand.* **18**, 1635 (1964); B. Beagley and R. W. H. Small, *Acta Crystallogr.* **17**, 783 (1964); D. J. Hodgson and J. A. Ibers, *Acta Crystallogr.* **B25**, 469 (1969).
(7) L. E. Sutton (ed.), *Chem. Soc. Spec. Publ.* No. 11 (1958); No. 18 (1965).
(8) C. H. Bibart and G. E. Ewing, *J. Chem. Phys.* **61**, 1284 (1974).
(9) R. G. Snyder and I. C. Hisatsune, *J. Mol. Spectrosc.* **1**, 139 (1957).
(10) I. C. Hisatsune, *J. Phys. Chem.* **65**, 2249 (1961).
(11) J. R. Durig, J. W. Thompson, J. D. Witt, and J. D. Odom, *J. Chem. Phys.* **58**, 5339 (1973).

(12) D. D. Danielson, J. V. Patton, and K. Hedberg, *J. Am. Chem. Soc.* **99**, 6484 (1977).

(13) N. J. Fitzpatrick, *Inorg. Nucl. Chem. Lett.* **9**, 965 (1973).

(14) M. F. Guest and I. H. Hillier, *J. Chem. Soc. Faraday II* **70**, 398 (1974).

(15) J. M. Howell and J. R. VanWazer, *J. Am. Chem. Soc.* **96**, 7902 (1974).

(16) L. Trefonas and W. N. Lipscomb, *J. Chem. Phys.* **28**, 54 (1958).

(17) J. R. Durig, J. E. Saunders, and J. D. Odom, *J. Chem. Phys.* **54**, 5285 (1971).

(18) M. Atoji, P. J. Wheatley, and W. N. Lipscomb, *J. Chem. Phys.* **27**, 196 (1957).

(19) K. Hedberg and R. R. Ryan, *J. Chem. Phys.* **41**, 2214 (1964).

(20) R. R. Ryan and K. Hedberg, *J. Chem. Phys.* **50**, 4986 (1969).

(21) L. H. Jones and R. R. Ryan, *J. Chem. Phys.* **57**, 1012 (1972).

(22) D. E. Mann and L. Fano, *J. Chem. Phys.* **26**, 1665 (1957).

(23) J. D. Odom, J. E. Saunders, and J. R. Durig, *J. Chem. Phys.* **56**, 1643 (1972).

(24) J. H. Robertson, *Acta Crystallogr.* **18**, 410 (1965).

(25) M. J. Schmelz, T. Miyazawa, S.-I. Mizushima, T. J. Lane, and J. V. Quagliano, *Spectrochim. Acta* **9**, 51 (1957).

(26) G. M. Begun and W. H. Fletcher, *Spectrochim. Acta* **19**, 1343 (1963).

(27) P. J. Wheatley, *J. Chem. Soc.* 4514 (1956).

(28) A. Snelson, B. N. Cyvin, and S. J. Cyvin, *Z. Anorg. Allg. Chem.* **410**, 206 (1974).

(29) A. Büchler and W. J. Klemperer, *J. Chem. Phys.* **29**, 121 (1958); G. A. Ozin, *Progr. Inorg. Chem.* **14**, 173 (1971).

(30) F. Seel, R. Budenz and W. Gombler, *Chem. Ber.* **103**, 1701 (1970); F. Seel and R. Budenz, *J. Fluorine Chem.* **1**, 117 (1971).

(31) J. D. Dunitz, *J. Am. Chem. Soc.* **78**, 878 (1956); J. D. Dunitz, *Acta Crystallogr.* **9**, 579 (1956).

(32) L. Burlamacchi, G. Casini, O. Fagioli, and E. Tiezzi, *Ric. Sci.* **37**, 97 (1967).

(33) H. L. Hodges, L. S. Su, and L. S. Bartell, *Inorg. Chem.* **14**, 599 (1975).

(34) C. R. S. Dean, A. Finch, P. J. Gardner, and D. W. Payling, *J. Chem. Soc. Faraday I*, **70**, 1921 (1974).

(35) Y. C. Leung and J. Waser, *J. Phys. Chem.* **60**, 539 (1956).

(36) G. Shanmugasundaram and G. Nagarajan, *Monatsh. Chem.* **100**, 789 (1969).

(37) A. V. Grosse, A. G. Streng and A. D. Kirshenbaum, *J. Am. Chem. Soc.* **83**, 1004 (1961); A. Arkell, *J. Am. Chem. Soc.* **87**, 4057 (1965); B. Plesnicar, D. Kocjan, S. Murovec and A. Azman, *J. Am. Chem. Soc.* **98**, 3143 (1976).

(38) C. J. Schack and D. Pilipovich, *Inorg. Chem.* **9**, 1387 (1970); K. O. Christe, C. J. Schack, and E. C. Curtis, *Inorg. Chem.* **10**, 1589 (1971).

(39) J.-L. Pascal and J. Potier, *Chem. Commun.* 446 (1973).

(40) C. H. W. Jones, *J. Chem. Phys.* **62**, 4343 (1975).

(41) W. E. Dasent and T. C. Waddington, *J. Chem. Soc.* 3350 (1960).

(42) S. N. Foner and R. L. Hudson, *J. Chem. Phys.* **58**, 581 (1973).

(43) K. J. Palmer and N. Elliot, *J. Am. Chem. Soc.* **60**, 1852 (1938).

(44) S. C. Wallwork and I. H. Worrall, *J. Chem. Soc.* 1816 (1965).

(45) B. Krebs, S. Pohl, and W. Schiwy, *Z. Anorg. Allg. Chem.* **393**, 241 (1972).

(46) J. W. Coomber and E. Whittle, *Trans. Faraday Soc.* **63**, 1394 (1967).

(47) J. A. Franklin and G. H. Hybrechts, *Int. J. Chem. Kinet.* **1**, 3 (1969).

(48) D. F. Eggers, Jr., R. C. Lord, and C. W. Wickstrom, *J. Mol. Spectrosc.* **59**, 63 (1976).

(49) D. A. Swick, I. L. Karle and J. Karle, *J. Chem. Phys.* **22**, 1242 (1954); Y. Morino, *J. Chem. Phys.* **28**, 185 (1958); J. Karle, *J. Chem. Phys.* **45**, 4149 (1966); G. Allen, P. N. Brier and G. Lane, *Trans. Faraday Soc.* **63**, 824 (1967).

(50) G. Heublein, R. Kühmstedt, P. Kadura, and H. Dawczynski, *Tetrahedron* **26**, 81 (1970).

(51) K. L. Brown and D. Hall, *J. Chem. Soc. Dalton* 1843 (1973).

(52) H. J. Cumming, D. Hall and C. E. Wright, *Cryst. Struct. Commun.* **3**, 103 (1974).

(53) A. Wilson and H. McGeachin, *Acta Crystallogr.* **17**, 1352 (1964).
(54) W. Klinger, G. Eulenberger, and H. Hahn, *Z. Anorg. Allg. Chem.* **401**, 97 (1973).
(55) K. L. Gallaher, A. Yokozeki, and S. H. Bauer, *J. Phys. Chem.* **78**, 2389 (1974).
(56) A. Almenningen, B. Andersen, and M. Traetteberg, *Acta Chem. Scand.* **18**, 603 (1964).
(57) G. Mandel and J. Donohue, *Acta Crystallogr.* **B28**, 1313 (1972).
(58) F. Höfler, W. Sawodny, and E. Hengge, *Spectrochim Acta* **26A**, 819 (1970); E. Hengge, *Monatsh. Chem.* **102**, 734 (1971); F. Höfler, S. Waldhör, and E. Hengge, *Spectrochim. Acta* **28A**, 29 (1972).
(59) B. H. Freeland, J. L. Hencher, D. G. Tuck, and J. G. Contreras, *Inorg. Chem.* **15**, 2144 (1976).
(60) R. A. Carney, E. A. Piotrowski, A. G. Meister, J. H. Braun, and F. F. Cleveland, *J. Mol. Spectrosc.* **7**, 209 (1961).
(61) D. W. H. Rankin and A. Robertson, *J. Mol. Struct.* **27**, 438 (1975); H. Oberhammer, *J. Mol. Struct.* **31**, 237 (1976).
(62) S. Martinez, S. Garcia-Blanco and L. Rivoir, *Acta Crystallogr.* **9**, 145 (1956); J. A. Rausell-Colom and S. Garcia-Blanco, *Acta Crystallogr.* **21**, 672 (1966); R. J. Guttormson and E. Stanley, *Acta Crystallogr.* **B25**, 971 (1969); R. N. Hargreaves and E. Stanley, *Z. Kristallogr.* **135**, 399 (1972).
(63) M. Bodenstein, P. Harteck, and E. Padelt, *Z. Anorg. Allg. Chem.* **147**, 233 (1925).
(64) C. J. Schack and K. O. Christe, *Inorg. Chem.* **13**, 2378 (1974).
(65) A. S. Parkes and R. E. Hughes, *Acta Crystallogr.* **16**, 734 (1963).
(66) M. G. Inghram, R. F. Porter, and W. A. Chupka, *J. Chem. Phys.* **25**, 498 (1956).
(67) K. M. Maloney, S. K. Gupta, and D. A. Lynch, Jr., *J. Inorg. Nucl. Chem.* **38**, 49 (1976).
(68) R. H. Jackson, *J. Chem. Soc.* 4585 (1962).
(69) R. L. Kuczkowski, *J. Am. Chem. Soc.* **86**, 3617 (1964).
(70) B. Beagley, G. H. Eckersley, D. P. Brown, and D. Tomlinson, *Trans. Faraday Soc.* **65**, 2300 (1969).
(71) E. Hirota, *Bull. Chem. Soc. Jpn.* **31**, 130 (1958).
(72) S. G. Frankiss, *J. Mol. Struct.* **2**, 271 (1968).
(73) G. Krummel and R. Minkwitz, *Inorg. Nucl. Chem. Lett.* **13**, 213 (1977).
(74) A. D. Kirshenbaum, A. V. Grosse, and J. G. Aston, *J. Am. Chem. Soc.* **81**, 6398 (1959).
(75) M. M. Rochkind and G. C. Pimentel, *J. Chem. Phys.* **46**, 448 (1967); W. G. Alcock and G. C. Pimentel, *J. Chem. Phys.* **48**, 2373 (1968); F. K. Chi and L. Andrews, *J. Phys. Chem.* **77**, 3062 (1973).
(76) R. J. Gillespie, J. B. Milne, and M. J. Morton, *Inorg. Chem.* **7**, 2221 (1968).
(77) W. Giggenbach, *J. Inorg. Nucl. Chem.* **30**, 3189 (1968).
(78) S. C. Abrahams, *Acta Crystallogr.* **7**, 423 (1954).
(79) D. D. Davis and H. Okabe, *J. Chem. Phys.* **49**, 5526 (1968); M. W. Slack, E. S. Fishburne, and A. R. Johnson, *J. Chem. Phys.* **54**, 1652 (1971).
(80) D. White, D. E. Mann, P. N. Walsh, and A. Sommer, *J. Chem. Phys.* **32**, 481 (1960).
(81) S. Smoes, C. E. Myers, and J. Drowart, *Chem. Phys. Lett.* **8**, 10 (1971).
(82) J. Drowart, R. P. Burns, G. DeMaria, and M. G. Inghram, *J. Chem. Phys.* **31**, 1131 (1959).
(83) K. S. Pitzer and E. Clementi, *J. Am. Chem. Soc.* **81**, 4477 (1959).

9 The physical underpinnings of qualitative MO theory

It is clear that qualitative MO arguments work for a wide range of molecular systems and molecular properties. There has been a great deal of effort devoted to the search for a fundamental understanding of why such a superficially simple method should be so successful. This problem has been discussed in detail by Allen (*1,2*), and more recently Buenker and Peyerimhoff (*3*) have published an exhaustive review. The reader should consult these references for an appreciation of the large number of examples that have been studied.

In qualitative MO theory, changes in molecular shape produce changes in the energies of individual electrons. These energy changes are then used to predict molecular conformations. Exactly what kind of one-electron energies are involved here? Are they the ionization potentials? Since molecular shape is determined by the minimum in the total molecular energy, must the individual electron energies sum to the total energy? If so, the use of ionization potentials is inappropriate since they do not have this property. What kind of one-electron energies do sum to the total energy?

CANONICAL ORBITS

The total SCF energy E can be written as

$$E = \sum \varepsilon_k + V_{nn} - V_{ee}, \qquad [1]$$

where V_{nn} and V_{ee} are, respectively, the nuclear-nuclear and electron-electron repulsive potential energies. The orbital energies ε_k are often called the canonical orbital energies because they come directly from the solution of the Hartree–Fock–Roothaan SCF equations (4) and they correspond to ionization potentials according to Koopmans' theorem. The summation

$$E' = \sum \varepsilon_k$$

is over all electrons in occupied MOs. For comparisons with qualitative MO theory this sum should be separated into parts corresponding to core and valence electrons:

$$E' = E'_{core} + E'_{val},$$

where E'_{core} is the sum over electrons in the core orbitals and E'_{val} is the sum over electrons in the valence orbitals. Then Eq. [1] can be rewritten as

$$E = E'_{val} + E'_{core} + V_{nn} - V_{ee}. \qquad [1a]$$

An immediate question is whether or not the canonical valence orbital energies ε_k are related to the orbital energies of qualitative MO theory. In other words, do the canonical orbital energies of ab initio SCF MO calculations reproduce the angular orbital correlation diagrams of qualitative MO theory? Allen and co-workers (5) have reported a great many calculations

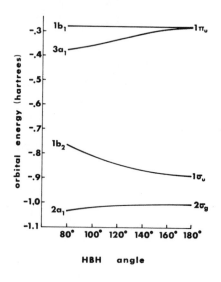

FIG. 1   Ab initio canonical orbital angular correlation diagram for BH_2^+. Orbital numbering includes core orbitals. (After Ref. 5.)

for small polyatomic molecules, the results demonstrating that the energies of occupied canonical valence orbitals undergo changes with valence angles that look just like those of the qualitatively deduced correlation diagrams. For example, compare the ab initio-based correlation diagram for BH_2^+ (Fig. 1) with the qualitative MO diagram for an AH_2 molecule (Fig. 1, Chapter 3).

Another important question concerns the universality of the calculated correlation diagrams. To what extent is a diagram that was calculated for one specific molecule similar to those calculated for other molecules of the same class either with the same or with different numbers of valence electrons? Figure 2 demonstrates the similar overall pattern of ab initio calculated correlation diagrams for BeH_3^-, BH_3, and CH_3^+ (5). Again, these diagrams are comparable to the qualitative MO picture for AH_3, as shown in Fig. 4, Chapter 3.

Is there a relationship between the total SCF energy E and the sum E'_{val} of occupied valence orbital energies? Figure 3 compares these two quantities as functions of valence angle for BH_2^-. Although E and E'_{val} are widely different numbers, they not only give the same qualitative shape conclusions but they also have minima at about the same valence angle. Consider the variation of energy with valence angle θ. For the two derivatives $\partial E/\partial \theta$ and $\partial E'_{val}/\partial \theta$ to vanish at the same equilibrium angle θ_e requires only that both derivatives have the same sign at all angles, a much less severe requirement than the equality of E and E'_{val} or the equality of their derivatives

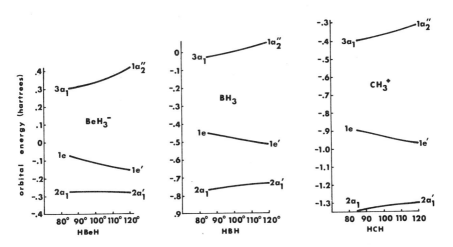

Fig. 2 Ab initio orbital angular correlation diagrams for BeH_3^-, BH_3, and CH_3^+. (After Ref. 5.)

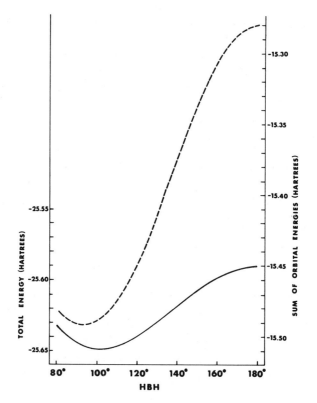

FIG. 3   Ab initio total energy E (——) and canonical orbital energy sum E'_{val} (– – –) as a function of valence angle for BH_2. (After Ref. 5.)

with respect to θ (1,2) (1). Taking the derivative of Eq. [1a] with respect to θ yields

$$\frac{\partial E}{\partial \theta} = \frac{\partial E'_{val}}{\partial \theta} + \frac{\partial}{\partial \theta} (E'_{core} + V_{nn} - V_{ee}).$$

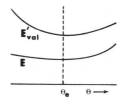

1

The quantities $\partial E'_{val}/\partial\theta$ and $\partial(E'_{core} + V_{nn} - V_{ee})/\partial\theta$ either have the same sign or different signs. If they have the same sign, then their ratio is positive:

$$\frac{\partial E'_{val}/\partial\theta}{\partial(E'_{core} + V_{nn} - V_{ee})/\partial\theta} \geq 0. \qquad [2]$$

If the two derivatives have different signs, then $\partial E'_{val}/\partial\theta$ must have the larger absolute value,

$$\left|\frac{\partial E'_{val}}{\partial\theta}\right| > \left|\frac{\partial}{\partial\theta}(E'_{core} + V_{nn} - V_{ee})\right|, \qquad [3]$$

in order to maintain the same sign between $\partial E/\partial\theta$ and $\partial E'_{val}/\partial\theta$. The relationships in Eqs. [2] and [3] suggest that the sum of valence orbital energies E'_{val} will break down as a predictor of molecular shape when the electrostatic potential energies V_{nn} and V_{ee} vary in different ways with changes in valence angle θ. This situation might be expected to occur when the electron cloud does not smoothly cover the nuclear framework, that is, in molecules with highly ionic bonding. The Li_2O molecule is an example. Both qualitative MO reasoning and the ab initio valence orbital sum E'_{val} predict that Li_2O should be bent like H_2O. The total SCF energy E, however, indicates that Li_2O has a linear structure, in agreement with experiment (6). Figure 4 shows E and E'_{val} for Li_2O. The fact that the qualitative model and the ab initio orbital sum E'_{val} both fail to give the correct shape for Li_2O adds support to the idea that the orbital energies of the qualitative model should be identified with the canonical MOs of ab initio SCF theory.

How are the extended Hückel orbital energies related to the ab initio SCF canonical orbital energies? Allen and Russell (7) have compared molecular shapes obtained from the E'_{val} minimum using extended Hückel and ab initio SCF calculations. For a large number of covalent molecules the agreement is satisfactory. Figure 5 compares E'_{val} for BH_2^- as calculated by extended Hückel and ab initio SCF methods. Even the shape prediction failure for Li_2O is reproduced by extended Hückel results. Still, the relationship between extended Hückel and ab initio canonical orbitals is not as simple as one might hope. For example, the extended Hückel method with suitable parametrizations is capable of giving gauche geometry for H_2O_2, but the sum of ab initio orbital energies does not.

The results of many calculations support, in a heuristic way, a relationship between qualitative MO theory, the extended Hückel method, and the canonical orbital energies of the SCF MO procedure. It is possible to show a formal, though still approximate, connection. The total energy E is the sum of kinetic (T) and potential (V) energy components:

$$E = T + V. \qquad [4]$$

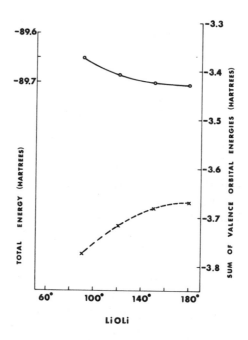

FIG. 4   Ab initio total energy E (——) and E'_{val} (---) versus valence angle for the ionic molecule Li_2O. (After Ref. 7.)

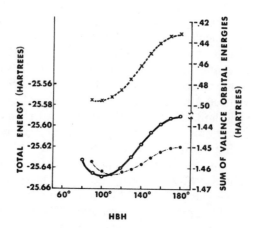

FIG. 5   Ab initio total energy E (——), ab initio E'_{val} (---), and extended Hückel E'_{val} (—·—·—·) versus valence angle for BH_2^-. (After Ref. 7.)

The total potential energy V can be partitioned into a sum of nuclear-electron attraction (V_{ne}), nuclear-nuclear repulsion (V_{nn}), and electron-electron repulsion (V_{ee}) terms:

$$V = V_{ne} + V_{nn} + V_{ee}.$$ [5]

The quantum mechanical virial theorem relates E, T, and V:

$$E = -T = +\tfrac{1}{2}V = \tfrac{1}{2}(V_{ne} + V_{nn} + V_{ee}).$$ [6]

Using approximations from the Thomas–Fermi treatment of atoms, Politzer and Parr (8) derived an equation relating the total energy of an atom and its nuclear-electron attraction potential energy:

$$E = \tfrac{3}{7}V_{ne}.$$ [7]

Fraga (9) had previously shown empirically the approximate validity of Eq. [7] for atomic SCF energies. Politzer (10) extended the arguments to obtain a comparable equation for the total molecular energy:

$$E = \tfrac{3}{7}(V_{ne} + 2V_{nn}).$$ [8]

Equation [8] is able to generate the total energies of over 50 molecules with an average error of 1.15% of the SCF energy using potentials V_{nn} and V_{ne} calculated from SCF wavefunctions.

Equation [1] gives the total SCF molecular energy as a sum of canonical orbital energies ε_k plus $V_{nn} - V_{ee}$. Ruedenberg (11) has recently shown that combining Eqs. [1], [6], and [8] leads to an expression for the total energy as a sum of canonical orbital energies only. Equating the energies in [6] and [8] yields

$$V_{ne} = 5V_{nn} - 7V_{ee}.$$

Substituting this relation for V_{ne} in Eq. [5] gives

$$\tfrac{1}{6}V = V_{nn} - V_{ee},$$ [9]

which clearly shows that V_{nn} and V_{ee} do not cancel in Eq. [1] and the total energy E is not directly a sum of the canonical orbital energies. From Eqs. [6] and [9] one obtains

$$\tfrac{1}{3}E = (V_{nn} - V_{ee}),$$

which, combined with Eq. [1], finally yields

$$E = \tfrac{3}{2}\sum \varepsilon_k.$$ [10]

Ruedenberg notes that Eq. [10] usually holds to within 2 to 4% for SCF MO calculations. Although Eq. [10] is approximate because Eq. [8] is approximate, the result does offer a formal relationship between the canonical orbital energies and the total energy of the system.

NONCANONICAL ORBITALS

A number of attempts have been made to express the total molecular energy E exactly as a sum of one-electron contributions or orbitals e_i,

$$E = \sum e_i \qquad [11]$$

in which no separate potential energy terms and no multiplicative factors appear. For example, Coulson and Neilson (12) defined the partitioned energies

$$e_i = \tfrac{1}{2}(\varepsilon_i + E_i),$$

where ε_i are the canonical orbitals and

$$E_i = \left\langle \phi_i \left| -\tfrac{1}{2}\nabla_1^2 - \sum_A \frac{Z_A}{r_{1A}} \right| \phi_i \right\rangle.$$

Then

$$E = \sum e_i + V_{nn}. \qquad [12]$$

Except for the nuclear-nuclear repulsion term, the e_i's in Eq. [12] sum directly to the total energy. Unfortunately, the partitioned energies e_i have neither the same energy ordering nor the same angular dependence as the qualitative or the canonical orbital energies. Compare the e_i's for water in Fig. 6 with the AH_2 correlation diagram, Fig. 1, Chapter 3.

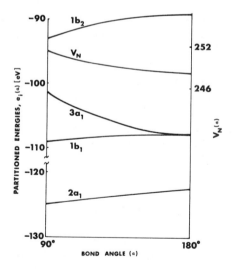

FIG. 6   Coulson and Nielson's partitioned energies, e_i versus bond angle for H_2O. (After Ref. 12.)

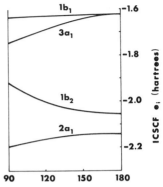

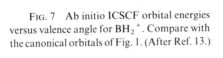

FIG. 7 Ab initio ICSCF orbital energies versus valence angle for BH_2^+. Compare with the canonical orbitals of Fig. 1. (After Ref. 13.)

Davidson (13) has noted an arbitrariness in the Fock operator of the Hartree–Fock–Roothan equations. He has used that flexibility to define a new Fock operator, the eigenvalues of which have the property that they sum directly to the total SCF energy as in Eq. [11]. The orbitals corresponding to these new one-electron energies e_i are called the internally consistent SCF (ICSCF) orbitals. Stenkamp and Davidson (14) have formed orbital correlation diagrams with the ICSCF orbitals and compared their diagrams with those constructed from the canonical orbital set for a large number of AH_2 and AH_3 molecules and ions plus HCN, CO_2, and Li_2O. Figure 7 is the ICSCF orbital correlation diagram for BH_2^+. Except for a different scale the ICSCF diagram of Fig. 7 is strikingly similar to the canonical orbital diagram for BH_2^+ in Fig. 1.

Since the ICSCF orbital sum exactly equals the total energy E, the minima in both quantities coincide even in cases involving highly polar or ionic bonding, such as Li_2O, for which the sum of canonical orbitals fails to predict the correct molecular shape. It is not surprising therefore that the canonical and ICSCF orbital correlation diagrams have significant differences for Li_2O. Although the e_i's succeed where the canonical ε_i's from ab initio, extended Hückel, and qualitative arguments fail, a qualitative interpretation of ICSCF orbital energy changes is lacking. The ICSCF orbital energies do not have the physical interpretation as ionization potentials enjoyed by the canonical orbital energies.

THE HELLMANN–FEYNMAN ELECTROSTATIC FORCE MODEL

In order to study a property of individual orbitals that sums directly to determine molecular shapes, Coulson and Deb (15) shifted their attention from the molecular energy to the electrostatic forces acting on the atoms in a molecule. They used the Hellmann–Feynman theorem to express the force F acting on a nucleus A in a molecule:

$$F_A = -\frac{\partial W}{\partial \mathbf{R}_A} = -\frac{\partial}{\partial \mathbf{R}_A} \langle \psi | H | \psi \rangle = -\left\langle \psi \left| \frac{\partial H}{\partial \mathbf{R}_A} \right| \psi \right\rangle, \qquad [13]$$

which expands to

$$F_A = Z_A \int \rho(\mathbf{r}_1) \frac{\mathbf{r}_{A1}}{r_{A1}^3} \, d\mathbf{r}_1 - Z_A \sum_{B \neq A} Z_B \frac{\mathbf{R}_{AB}}{R_{AB}^3}, \qquad [14]$$

where the first term (the integral) represents attractions between electrons and nucleus A of charge Z_A and the second term (the sum) is the total of internuclear repulsions between A and all other nuclei B. If the wavefunction ψ is approximated by a molecular orbital wavefunction, then the one-electron density function $\rho(r)$ can be expressed as a sum of contributions from individual MOs.

Deb (16) has used this idea to show qualitatively how electrostatic forces determine molecular shapes. Those angular geometry changes that increase the overlap between AOs on A and B in a MO increase the electron density function ρ and, therefore, the attractive force acting on the nucleus. Pictures showing qualitatively the AO composition of each MO, particularly the highest occupied MOs, can then be used to determine which angular variations produce increased attractive forces. In practice, this scheme works exactly like the qualitative orbital energy method. Both require a set of MOs from which molecular shape conclusions are drawn by interpreting changes in AO overlaps. In the electrostatic force theory, nuclear-electron attractive forces increase as AO overlaps increase, while in the qualitative orbital energy model AO overlap increases lower the MO energy. The transformation from energies to forces does not, therefore, lead to a more fundamental interpretation of the success of the qualitative MO model. In fact, the approximations and physical origins of the electrostatic force model and the qualitative MO model are identical. The choice between the two schemes might thus appear to be a matter of taste, but keep in mind that orbital energies are an immediate product of any MO calculation and, compared to forces, energies are more familiar and useful quantities for discussions of other properties such as bond energies, reaction mechanisms, and relative molecular stabilities.

In a fascinating series of articles, Nakatsuji (17) has extended the Hellmann–Feynman electrostatic force theory both in formalism and as a practical method, linking its concepts to those of traditional valence theory.

ANGULAR GEOMETRY AND INFORMATION CONTENT

Allen (1,2) has characterized the angular geometry or shape property of molecules as having low information content. Qualitative MO arguments and extended Hückel calculations are sufficient to predict the shapes of most small molecules. Compared to ab initio techniques, the qualitative and semiempirical methods employ only a modest amount of input information and generate far less in the way of intermediate results such as the values of molecular integrals required for the final results. If the simple schemes are successful in determining a particular molecular property, then the property in question must have a low information content. Although molecular shape has low information content, the same is not necessarily true of other properties. Accurate bond distances, for instance, cannot be obtained successfully by minimizing sums of orbital energies from either ab initio or extended Hückel calculations. The collapse of H_2 into the united atom He as predicted by qualitative MO theory is another example that qualitative arguments applied to bond distances must be made with caution.

RELATIONSHIP BETWEEN THE QUALITATIVE MO AND THE VSEPR MODELS

The most familiar qualitative model for molecular structure predictions is the valence shell electron pair repulsion (VSEPR) model of Sidgwick and Powell (18) and Gillespie and Nyholm (19). Gillespie's book provides an up-to-date and comprehensive discussion of the model and its applications (20). The VSEPR model is particularly easy to use because it involves a single, usually well-defined valence bond (VB) electronic structure picture and requires concepts no more complicated than electrostatic and steric repulsions. Very large molecules are treated just as easily as small ones. However, not all molecular conformations result from repulsive forces. Furthermore, the VSEPR model cannot be conveniently applied to excited states of molecules nor does it provide a picture of the limitations on reaction mechanisms comparable to the conservation of orbital symmetry approach of qualitative MO theory. Finally, the ideas of delocalized electrons and multicenter bonding must be grafted onto the VSEPR model in order to account for mobile pi bonds such as those in benzene or ozone and to explain electron deficient bonds such as those in the hydrogen bridges of the boron hydrides.

It has long been known that a relationship exists between the delocalized canonical orbitals of MO theory and the localized bond and lone pair orbitals of the VB model. A unitary transformation $\overline{U}$ can transform the set of occupied, delocalized MOs $\bar{\phi} = \{\phi_i\}$ into a set of occupied, localized bond and lone pair orbitals $\bar{\chi}$, without changing the total electron distribution,

$$\bar{U}\bar{\phi} = \bar{\chi}.$$

Over the years many different criteria for localization have been proposed. A choice with considerable physical appeal is the one described by Edmiston and Ruedenberg (21) who have shown how to obtain the $\bar{U}$ that minimizes the overall interorbital Coulombic repulsion part of the total electron-electron repulsion energy V_{ee}. For molecules such as F_2 and HF, the localized orbitals produced by this method are very similar to those one would choose on the basis of chemical intuition.

The object here, however, is to compare the qualitative forms of the MO and VB theories. The result of the unitary transformation $\bar{U}$ is to form new functions as linear combinations of the old. Therefore, the localized orbitals can be formed by combining two or more delocalized MO pictures. For example, equivalent bond functions for each O-H bond in the water molecule can be made from the sum and difference of the bonding MOs $2a_1$ (before mixing) and $1b_2$ (2). Occasionally, a localized orbital turns out

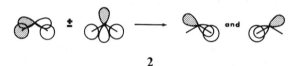

2

to be the same as an individual MO. The ammonia lone pair orbital $2a_1$ (after mixing) is one example; several other cases occur in the earlier chapters.

The orbital transformation process is, of course, reversible; Thompson and Davidson have shown how delocalized MOs can be produced by mixing together intuitively formed localized bond orbitals (22).

REFERENCES

(1) L. C. Allen, Why three dimensional Hückel theory works and where it breaks down, *in* "Sigma Molecular Orbital Theory" (O. Sinanoglu and K. B. Wiberg, eds.), p. 227. Yale Univ. Press, New Haven, Connecticut, 1970.

(2) L. C. Allen, *Theoret. Chim. Acta* **24**, 117 (1972).

(3) R. J. Buenker and S. D. Peyerimhoff, *Chem. Rev.* **74**, 127 (1974).

(4) C. C. J. Roothaan, *Rev. Mod. Phys.* **23**, 161 (1951).

(5) BH_2^+, BH_2^-, NH_2^+, BeH_2, BH_3, CH_3^+, BeH_3^-: S. D. Peyerimhoff, R. J. Buenker, and L. C. Allen, *J. Chem. Phys.* **45**, 734 (1966); NH_3: A. Rauk, L. C. Allen, and E. Clementi, *J. Chem. Phys.* **52**, 4133 (1970); PH_3, PH_5: A. Rauk, L. C. Allen and K. Mislow, *J. Am. Chem. Soc.* **94**, 3035 (1972); BH_3, BF_3: M. E. Schwartz and L. C. Allen, *J. Am. Chem. Soc.* **92**,

1466 (1970); HCN: D. C. Pan and L. C. Allen, *J. Chem. Phys.* **46**, 1797 (1967); F_2O, Li_2O, FOH, LiOH: R. J. Buenker and S. D. Peyerimhoff, *J. Chem. Phys.* **45**, 3682 (1966); CO_2, BeF_2: S. D. Peyerimhoff, R. J. Buenker, and J. L. Whitten, *J. Chem. Phys.* **46**, 1707 (1967); O_3, N_3^-: S. D. Peyerimhoff and R. J. Buenker, *J. Chem. Phys.* **47**, 1953 (1967); N_2O: S. D. Peyerimhoff and R. J. Buenker, *J. Chem. Phys.* **49**, 2473 (1968); NO_2^+, NO_2^-: G. V. Pfeiffer and L. C. Allen, *J. Chem. Phys.* **51**, 190 (1969); C_2H_2, C_2H_4: R. J. Buenker, S. D. Peyerimhoff, and J. L. Whitten, *J. Chem. Phys.* **46**, 2029 (1967); H_2O_2, CH_3OH, CH_3NH_2: W. H. Fink and L. C. Allen, *J. Chem. Phys.* **46**, 2261, 2276 (1967); N_2H_2: D. P. Wong, W. H. Fink, and L. C. Allen, *J. Chem. Phys.* **47**, 895 (1967); B_2H_6, C_2H_6: R. J. Buenker, S. D. Peyerimhoff, J. L. Whitten, and L. C. Allen, *J. Chem. Phys.* **45**, 2835 (1966).

(6) A. Büchler, J. L. Stauffer, W. Klemperer, and L. Wharton, *J. Chem. Phys.* **39**, 2299 (1963); D. White, K. S. Seshadri, D. F. Dever, D. E. Mann, and M. J. Linevsky, *J. Chem. Phys.* **39**, 2463 (1963).

(7) L. C. Allen and J. D. Russell, *J. Chem. Phys.* **46**, 1029 (1967).

(8) P. Politzer and R. G. Parr, *J. Chem. Phys.* **61**, 4258 (1974).

(9) S. Fraga, *Theoret. Chim. Acta* **2**, 406 (1964).

(10) P. Politzer, *J. Chem. Phys.* **64**, 4239 (1976).

(11) K. Ruedenberg, *J. Chem. Phys.* **66**, 375 (1977).

(12) C. A. Coulson and A. H. Neilson, *Disc. Faraday Soc.* **35**, 71 (1963).

(13) E. R. Davidson, *J. Chem. Phys.* **57**, 1999 (1972).

(14) L. Z. Stenkamp and E. R. Davidson, *Theoret. Chim. Acta* **30**, 283 (1973).

(15) C. A. Coulson and B. M. Deb, *Int. J. Quantum Chem.* **5**, 411 (1971).

(16) B. M. Deb, *J. Am. Chem. Soc.* **96**, 2030 (1974); B. M. Deb, P. N. Sen, and S. K. Bose, *J. Am. Chem. Soc.* **96**, 2044 (1974).

(17) H. Nakatsuji, *J. Am. Chem. Soc.* **95**, 345, 354, 2084 (1973); **96**, 24 (1974); H. Nakatsuji, T. Kuwata, and A. Yoshida, *J. Am. Chem. Soc.* **95**, 6894 (1973); H. Nakatsuji and T. Koga, *J. Am. Chem. Soc.* **96**, 6000 (1974).

(18) N. V. Sidgwick and H. E. Powell, *Proc. Roy. Soc. London* **A176**, 153 (1940).

(19) R. J. Gillespie and R. S. Nyholm, *Quart. Rev.* **11**, 339 (1957).

(20) R. J. Gillespie, "Molecular Geometry." Van Nostrand-Reinhold, Princeton, New Jersey, 1972.

(21) C. Edmiston and K. Ruedenberg, *Rev. Mod. Phys.* **35**, 457 (1963); *J. Chem. Phys.* **43**, 597 (1965).

(22) H. B. Thompson, *Inorg. Chem.* **7**, 604 (1968); R. B. Davidson, *J. Chem. Educ.* **54**, 531 (1977).

10 The development of qualitative MO theory

THE PAST

The first diagram showing changes in MO energies with variations in angular geometry was drawn for AB_2 molecules by Mulliken in 1942 (*1*). Mulliken deduced his orbital correlation diagram from known molecular geometries, observed visible and ultraviolet spectra and ionization potentials, and qualitative MO ideas. This approach was consolidated by Walsh who published a classic series of papers in 1953 (*2*) in which he surveyed the shapes of molecules of several classes, such as AH_2, AH_3, and HAB, in a way that set the pattern for the chapters of this book. Each discussion centered on an angular orbital correlation diagram deduced from known molecular geometries, isoelectronic analogies, electronic spectra and ionization potentials, and qualitative MO and hybridization arguments. MO angular correlation diagrams of the type that Walsh and Mulliken published are now widely known as Walsh diagrams, or less commonly, Walsh–Mulliken diagrams. The summaries of observations that the shapes of molecules in a class are determined by the number of valence electrons are often called Walsh's rules.

Since the early 1950s many additional molecular structures have been determined by the well-established methods of X-ray and electron diffraction and microwave spectroscopy. Infrared, Raman, and electron spin resonance spectroscopies, often coupled with the techniques of matrix isolation, flash photolysis, and high-energy irradiation, have revealed the shapes of many unstable or reactive species that fill the gaps in large classes of molecules. Thus, there exists today a wealth of experimental structural detail that was not available to Walsh during his pioneering effort 25 years ago.

During the 1960s great advances were made in quantitative MO methods, both at the ab initio SCF and the semiempirical levels. The results of the ab initio calculations have shown that most of chemistry is contained within the Hartree–Fock limit of MO theory. In particular, geometries of small molecules can be calculated with bond angles to within a few degrees and bond distances to within a few hundredths of an Angstrom of the best experimental values. In fact, for many difficult to observe species, geometries determined by ab initio SCF MO calculations are often far more reliable than those obtained from experiment. Because the MO calculations are relatively convenient to execute and can be carried out to comparable levels of approximation for many different classes of molecules, the use of MO wavefunctions rapidly surpassed that of valence bond wavefunctions in valence theory calculations.

The formulation of the semiempirical, semiquantitative extended Hückel MO method as a practical computational scheme by Hoffmann in 1963 was significant for the development of qualitative MO theory (3). Hoffmann's scheme and similar non-SCF methods are sometimes referred to as generalized Hückel or three-dimensional Hückel methods. The extended Hückel calculations employ empirical AO energies as basic parameters, yet the method itself can be formally related to ab initio MO theory and thereby to the laws of quantum mechanics that underlie chemistry. Thus, the extended Hückel method serves to link fundamental physical principles and the data for atoms, which are the building blocks of molecules. The real value of the extended Hückel method is not in its quantitative results, which have never been impressive, but rather in the qualitative nature of the results and in the interpretations those results can provide. Since the extended Hückel calculations could be readily performed on electronic computers of modest capacity and speed, it became practical to accumulate orbitals and orbital energies for whole classes of molecules in various conformations and to test qualitative arguments made within the MO framework.

The most notable achievement of qualitative MO theory during the 1960s was the use of orbital symmetry considerations to rationalize the stereochemical pathways of concerted reactions. Woodward and Hoffmann described reaction mechanisms as being controlled by the principle of conservation of orbital symmetry (4,5). Independently and at about the same time, Fukui advanced very similar ideas, proposing that reaction paths are determined by maximum overlapping of the highest occupied and the lowest unoccupied MOs of the reacting species (6,7). Their success in using qualitative MO ideas to understand a large number of reaction mechanisms of organic chemistry stimulated renewed interest in qualitative MO theory at a time when experiment and both rigorous and semiempirical MO calculations provided a particularly rich accumulation of information for interpretation.

The bulk of the material in this book has been drawn from a series of papers published since 1970 (8–11). The major feature of this work was the interpretation of Walsh–Mulliken diagrams in terms of qualitative AO composition pictures, symmetry considerations, and overlap arguments. However, previous books have made similar connections. Works by Herzberg (12) and by Atkins and Symons (13) contain AO composition diagrams for the AH_2, AH_3, AB_2, and AB_3 classes. Burdett has presented Walsh–Mulliken diagrams for AB_n ($n = 2$–7) main group systems (14). The book by Salem and Jorgensen (15) features pictures of valence orbitals for a large number of familar covalent molecules. The pictures in the Salem and Jorgensen book are computer-drawn contour diagrams of ab initio SCF-generated orbitals and therefore they are much more accurate representations of the MOs than the qualitative AO composition diagrams included in this book. The similarity between the qualitative pictures and those obtained by rigorous methods is gratifying, but sometimes less is more. The qualitative pictures give a better idea of the AO composition of MOs and they usually make orbital energy changes and energy-level ordering easier to interpret. In a recent book, Pearson has used symmetry arguments and qualitative MO theory to predict whether the activation barrier for a given reaction path is likely to be large or small, taking examples from both organic and inorganic chemistry (16). Pearson's approach requires a knowledge of the energy order of molecular orbitals and their symmetries. Molecular shape predictions are based primarily on the second order Jahn–Teller effect.

THE FUTURE

The development and application of qualitative MO theory are not completed topics. Many classes of small molecules remain to be studied. This book has been restricted to s and p AO chemistry. Classes of molecules in which d AOs act as valence orbitals have been omitted. Hoffmann and co-workers have recently made a number of important qualitative MO studies that feature properties of transition element complexes (17), including d AOs in the basis set.

Only a limited group of molecular properties have been treated in this book. There has been no discussion of electronic spectra and ionization potentials, although Mulliken and Walsh made extensive use of these quantities in constructing their original orbital angular correlation diagrams. This book makes no mention of electron affinities although Lowe has recently demonstrated that qualitative MO theory can systematize trends in electron affinities (18). Much more work remains to be done to rationalize relative reaction rates and trends in molecular stabilities, particularly for inorganic species.

How can qualitative MO arguments be extended to large molecules? Obviously, the same concepts apply but difficulties rapidly increase as more atoms add more AOs to the basis set creating more MOs that lie closer together in energy. Because of the close energies, mixing among orbitals of the same symmetry is likely to be extensive, possibly causing a reordering of orbital energies. The uncertainty in energy-level ordering presents no serious problems unless highest occupied and lowest unoccupied orbitals are reversed. Several paths may be useful for the extension of qualitative MO theory to larger systems. First, it is possible to form the MOs of large molecules by taking appropriate combinations of MOs of small molecules as was done for A_2H_4, A_2H_6, A_2B_4, and A_2B_6. Another approach is to obtain only those MOs of chemical interest by transforming them from the valence bond or localized orbital structures as proposed by Thompson (19) and carried out by Hoffmann in numerous studies (20). Finally, the qualitative model can be used to interpret MOs generated by quantitative MO methods for large molecules.

REFERENCES

(1) R. S. Mulliken, *Rev. Mod. Phys.* **14**, 204 (1942).
(2) A. D. Walsh, *J. Chem. Soc.* 2260, 2266, 2288, 2296, 2301, 2306, 2318, 2321, 2325, 2330 (1953).
(3) R. Hoffmann, *J. Chem. Phys.* **39**, 1397 (1963).
(4) R. Hoffmann and R. B. Woodward, *Accounts Chem. Res.* **1**, 17 (1968).
(5) R. B. Woodward and R. Hoffmann, "The Conservation of Orbital Symmetry." Academic Press, New York, 1970.
(6) K. Fukui, "Theory of Orientation and Stereoselection." Springer-Verlag, Berlin and New York, 1970.
(7) K. Fukui, *Accounts Chem. Res.* **4**, 57 (1971).
(8) B. M. Gimarc, *Accounts Chem. Res.* **7**, 384 (1974); *J. Chem. Phys.* **53**, 1623 (1970); *J. Am. Chem. Soc.* **92**, 266 (1970); **93**, 593, 815 (1971); **95**, 1417 (1973); **100**, 2346 (1978); *Tetrahedron Lett.* **22**, 1859 (1977); B. M. Gimarc, J. F. Liebman, and M. C. Kohn, *J. Am. Chem. Soc.* **100**, 2334 (1978); B. M. Gimarc and S. A. Khan, *J. Am. Chem. Soc.* **100**, 2340 (1978); B. M. Gimarc, S. A. Khan, and M. C. Kohn, *J. Am. Chem. Soc.* **100**, 1996 (1978); J. R. Durig, B. M. Gimarc, and J. D. Odom, *Vib. Spect. Struct.* **2**, 1 (1973).
(9) R. Hoffmann, J. M. Howell, and E. L. Muetterties, *J. Am. Chem. Soc.* **94**, 3047 (1972); M. M. L. Chen and R. Hoffmann, *J. Am. Chem. Soc.* **98**, 1647 (1976); R. Hoffmann, J. M. Howell, and A. R. Rossi, *J. Am. Chem. Soc.* **98**, 2484 (1976).
(10) R. Gleiter and A. Veillard, *Chem. Phys. Lett.* **37**, 33 (1976).

(11) J. P. Lowe, *J. Am. Chem. Soc.* **92**, 3799 (1970).

(12) G. Herzberg, "Molecular Spectra and Molecular Structure" Vol. III, Electronic Spectra and Electronic Structure of Polyatomic Molecules. Van Nostrand-Reinhold, Princeton, New Jersey, 1966.

(13) P. W. Atkins and M. C. R. Symons, "The Structure of Inorganic Radicals: an Application of Electron Spin Resonance to the Study of Molecular Structure." Elsevier, Amsterdam, 1967.

(14) J. K. Burdett, *Struct. and Bonding* **31**, 67 (1976).

(15) L. Salem and W. L. Jorgensen, "The Organic Chemist's Book of Orbitals." Academic Press, New York, 1973.

(16) R. G. Pearson, "Symmetry Rules for Chemical Reactions." Wiley, New York, 1976.

(17) N. Rösch and R. Hoffmann, *Inorg. Chem.* **13**, 2656 (1974); R. Hoffmann, M. M. L. Chen, M. Elian, A. R. Rossi, and D. M. P. Mingos, *Inorg. Chem.* **13**, 2666 (1974); A. R. Rossi and R. Hoffmann, *Inorg. Chem.* **14**, 365 (1975); M. Elian and R. Hoffmann, *Inorg. Chem.* **14**, 1058 (1975); M. Eilan, M. M. L. Chen, D. M. P. Mingos, and R. Hoffmann, *Inorg. Chem.* **15**, 1148 (1976); P. J. Hay, J. C. Thibeault, and R. Hoffmann, *J. Am. Chem. Soc.* **97**, 4884 (1975); J. W. Lauher and R. Hoffmann, *J. Am. Chem. Soc.* **98**, 1729 (1976); J. W. Lauher, M. Elian, R. H. Summerville, and R. Hoffmann, *J. Am. Chem. Soc.* **98**, 3219 (1976); T. A. Albright, P. Hofmann, and R. Hoffmann, *J. Am. Chem. Soc.* **99**, 7546 (1977).

(18) J. P. Lowe, *J. Am. Chem. Soc.* **99**, 5557 (1977).

(19) H. B. Thompson, *Inorg. Chem.* **7**, 604 (1968).

(20) R. Hoffmann, *Accounts Chem. Res.* **4**, 1 (1971).

INDEX